The Power Within:
A Deep Dive into
Skeletal Muscle Function

Marcel

Table of Contents

Chapter 1. Introduction

1.1 Skeletal muscle biology

Skeletal muscles are largely comprised of skeletal myofibers, which are long, multinucleated cells that become bundled into fascicles within skeletal muscle tissue (Fig. 1A). Each myofiber is packed with myofibrils containing series of sarcomeres, the contractile unit of muscle. Each sarcomere consists of interdigitated strands of actin (thin) filament and myosin (thick) filament between two Z-discs of dense actin. During contraction, the myosin head units use ATP to ratchet along the actin filaments, shortening the sarcomere. Each actin filament is covered with strands of troponin and tropomyosin which function to block myosin-actin interaction during muscle relaxation. The sarcoplasmic reticulum of myofibers is a specialized form of endoplasmic reticulum, continuous with the extracellular space, and containing a reservoir of Ca^{2+} ions. Following depolarization of the myofiber membrane at the neuromuscular junction (NMJ), voltage-gated channels release Ca^{2+} into the cytoplasm from the extracellular space and the sarcoplasmic reticulum. Released Ca^{2+} ions bind tropomyosin causing a conformational change, revealing the myosin-binding sites on actin and allowing muscle contraction.

Satellite cells are the predominant stem cells in skeletal muscle. They have a high nucleus:cytoplasm ratio, and are tightly associated with myofibers and sandwiched between the sarcolemma and the myofiber-enveloping basal lamina.[1] Unlike myofibers, satellite cells are mitotically active and may proliferate, particularly during muscle growth or in response to muscle injury.[2-7] Satellite cells may also divide by asymmetric division to create myoblasts which subsequently fuse with growing or regenerating myofibers or create new myofibers in regions with previous damage.[8]

Muscle also contains fibroblasts and immune cells. Fibroblasts are loosely-defined, most often spindle-shaped cells that are responsible for producing the majority of the extracellular matrix (ECM) within muscle.[9-11] Macrophages and dendritic cells are the resident immune cells of muscle, involved in cleaning debris and surveying the tissue for pathogens. Neutrophils and T-cells transiently enter muscle in response to injury or the detection of a pathogen.[12]

1.2 The muscle-extracellular matrix interface

Skeletal muscle contains multiple layers of ECM. The endomysial ECM surrounds each myofiber, and includes the basal lamina and the reticular lamina. The basal lamina is composed primarily of laminin and collagen IV, while the reticular lamina contains collagen VI. The endomysial ECM also contains collagens I and III. The perimysial ECM bundles myofibers into fascicles, is continuous with the tendon, and is composed primarily of collagen I. The epimysium, again containing collagens I and III, is the outermost covering, surrounding the entire skeletal muscle. The ECM is crucial for muscle structural stability and the transmission of force between myofibers.[13, 14] The contracting sarcomere generates force within myofibers, which is then transmitted through dystrophin and dystroglycan to the muscle-ECM interface. All muscle ECM may also contain a variety of other glycoproteins, including proteoglycans, such as agrin, as well as matrix metalloproteinases, certain muscle trophic factors, and a large number of other laminin and/or collagen proteins.

1.2.1 Dystroglycan

Dystroglycan is a transmembrane glycoprotein translated as a single protein unit, but cleaved by autoproteolysis into α- and β-subunits (Fig. 1B).[15-17] Although the subunits subsequently bind tightly to each other through non-covalent binding via the α-dystroglycan C-terminus and the β-dystroglycan N-terminus,[18-22] the initial cleavage is nonetheless required for normal muscle development.[23] On the cytoplasmic face of the plasma membrane, the β-dystroglycan C-terminus binds the dystrophin C-terminus.[24-28] Additionally, β-dystroglycan binds inner-leaflet signaling proteins of the PI3K/Akt/mTOR pathway.[25, 29-36] Furthermore, recent studies found that β-dystroglycan can translocate to the nucleus where it regulates important nuclear functions and nuclear morphology.[37-39] Ervasti and Campbell first showed that glycosylation dictates laminin-binding to dystroglycan in 1993.[40] α-Dystroglycan is heavily glycosylated by a series of glycosyltransferase enzymes. Muscle ECM proteins, including perlecan, agrin[41, 42] and laminin[40] bind α-dystroglycan to link the ECM to the muscle sarcolemmal membrane.[40, 43, 44]

1.2.2 Integrins

There are 24 known heterodimers of integrin made from 18 α-subunit and 8 β-subunits.[45] Integrins engage in bidirectional signaling: typical "outside-in" signaling in which an extracellular ligand binds integrin and triggers and intracellular response, and "inside-out" signaling in which internal cellular processes modify integrins to regulate cell adhesion to the ECM. Like dystroglycan, integrins serve to link the ECM outside skeletal myofibers to the F-actin cytoskeleton inside muscle cells. In skeletal muscle, integrin-α7β1 is the predominant and most important integrin-laminin receptor; laminin-integrin interactions also trigger signals involved in cell survival and muscle regeneration.[46]

1.2.3 Laminins

Laminins are cruciform heterotrimeric proteins composed of α, β, and γ subunits. Laminins are named by the sequence of subunit isoforms, e.g. laminin-211 is laminin-α2, β1, and γ1.[47] The subunits interact in a coiled-coil domain, forming the long-arm of the cross. Different laminins exist within skeletal muscle, particularly at the NMJ, but the primary laminin in skeletal muscle ECM is laminin-211.[48, 49] The three N-termini of individual laminin-211 heterotrimers interact to achieve laminin-211 polymerization.[50-55] Within the basal lamina, laminin-211 is connected to collagen IV via a bridging nidogen protein.[53, 54, 56] The laminin-α subunits are unique in having five C-terminal globular (G) domains. The G domains bind membrane receptors dystroglycan[40, 43, 44] and integrins,[57, 58] integrin-α7β1 in skeletal muscle. Through binding membrane receptors and components of the collagen network, laminin-211 is crucial in anchoring myofibers to the ECM.

1.3 Mutations in *LAMA2* cause MDC1A

1.3.1 Genetics

Mutations in the *LAMA2* gene encoding laminin-α2 cause type 1A congenital muscular dystrophy (MDC1A).[59, 60] Causative mutations are typically frameshifts resulting in premature translation termination.[61-64] Loss of laminin-α2, and therefore of laminin-211, allows contracting myofibers to disconnect from the ECM whereupon they undergo apoptosis resulting in a severe and progressive muscular dystrophy.[65] Less commonly, point mutations may disrupt certain laminin-211 functions, such as polymerization, and cause a milder form of dystrophy. [66-68]

1.3.2 Human disease

MDC1A is a rare disease that typically presents in infancy, with neonatal hypotonia and joint contractures.[69] MDC1A can be distinguished from other congenital muscular dystrophies, and diagnosed by, genetic testing or by a reduction or absence of laminin-α2 staining on muscle biopsy.[63, 64] The disease progresses with axial and proximal muscle weakness, scoliosis or kyphosis, delayed motor development, loss of ambulation, and areflexia. Increased serum creatine kinase may be found on blood draw. MDC1A often results in premature death due to respiratory muscle weakness and associated infections.[70] On muscle biopsy, laminin-α2-deficient muscle displays typical signs of muscular dystrophy including necrosis, inflammation, and reduced myofiber size. Laminin-α2-deficient muscle also has signs of degeneration-regeneration cycles: increased variation in myofiber size and central nucleation.[71] Furthermore, laminin-α2-deficient muscle may display compensatory upregulation of laminin-α4 protein.[72]

1.3.3 Animal Models

Several MDC1A mouse models exist and display different disease severity.[73] The dystrophia muscularis (dy) mouse was the first mouse model, containing a natural mutation in the *Lama2* gene.[74] Subsequent mouse models were created by interrupting the *Lama2* gene with an inserted neomycin resistance gene (dy[3K]) or a *LacZ* gene (dy[W]).[75, 76] Although disease progression and lifespan varies in dy[W] mice,[77] and they retain some residual expression of a truncated laminin-α2,[73] they remain a frequently used model of MDC1A. Dy[W] mice typically begin showing signs of muscular dystrophy at 2 - 3 weeks of age, including smaller body weight and muscle weakness. Dy[W] muscle displays necrosis, fibrosis, reduced myofiber size, increased myofiber size variation, and increased central nucleation.[76] Additionally, dy[W] mice display reduced muscle regeneration.[78]

1.4 Mutations in dystroglycan glycosylation genes cause dystroglycanopathies

1.4.1 Genetics

Loss of function mutations in *DAG1*, encoding dystroglycan are lethal in mice (and likely lethal in humans), however point mutations affecting α-dystroglycan glycosylation may cause a primary dystroglycanopathy.[79] Secondary dystroglycanopathies, or α-dystroglycanopathies, are more common than primary dystroglycanopathies. These diseases may be caused by mutations in any one of the 24 or more genes involved in glycosylating α-dystroglycan, and thereby mediating its interaction with laminin-211. The enzymes encoded by these genes vary from glycosyltransferases such as protein-O-mannosyl transferase (POMT1), to kinases such as protein-O-mannose kinase (POMK), to catalysts for sugar generation such as isoprenoid synthase domain containing (ISPD).[80-82] Likewise, dystroglycanopathies vary from mild to severe, however, there is not a clear correlation between genotype and disease severity at this time.[83]

1.4.2 Human disease

Like MDC1A, severe cases of dystroglycanopathy present in infancy with neonatal hypotonia, joint contractures, and feeding difficulties.[69] Dystrophy may be mild to severe. In addition, the dystroglycanopathies often involve brain and eye anomalies. On muscle biopsy, these diseases typically display reduced binding of the IIH6 antibody, specific for α-dystroglycan's laminin-binding domain.[80, 81] Mutations in fukutin-related protein (FKRP) or like-acetylglucosaminyltransferase (LARGE) may result in the severe or mild forms of dystroglycanopathy. Milder disease mutations result in forms of limb-girdle muscular dystrophy. Other mutations result in a range of overlapping severe neonatal diseases including Walker-Warburg syndrome, muscle-eye-brain disease (MEB), and Fukuyama congenital muscular dystrophy (FCMD). In addition to muscular dystrophy, these diseases may involve brain anomalies, including cobblestone lissencephaly, agyria, hydrocephalus, neuronal heterotopia, cerebellar hypoplasia, incomplete or absent corpus callosum, and absent septum. Depending on brain involvement, motor and intellectual disability may be severe. Though dystroglycanopathies are generally rare, increased incidence of MEB occurs in Finland due to a founder mutation in *POMGNT1*, and FCMD is the second most common type of muscular dystrophy in Japan, behind Duchenne muscular dystrophy.[81]

1.4.3 Animal Models

Many animal models have been created to study dystroglycanopathy and the role of dystroglycan in muscle and brain development. While *Dag1*-null mice die in utero due to failure of Reichert's membrane,[84] tissue-specific *Dag1* knock-out mice have been studied.[85-87] Additional models include expression of cleavage-resistant dystroglycan,[88] partial loss of β-dystroglycan,[86] POMGnT1,[89, 90] POMGnT2,[91] fukutin,[92, 93] FKRP,[93-97] and Large-mutated mice.[86, 98-101] The Large[vls] mouse is a recently discovered line with a deletion in *Large* causing a frameshift and premature STOP codon. These mice display skeletal muscle pathology including fibrosis, collagen deposition, and fatty replacement of muscle tissue. They also have cardiac muscle pathology including degeneration, necrosis, fibrosis, and adult-onset cardiomyopathy. While Large[vls] mice do exhibit developmental defects of the eyes, they display little central nervous system involvement, potentially making them good subjects for studying the isolated muscular aspects of dystroglycanopathy.[102]

1.5 Current state of therapies

There are currently no FDA-approved therapies for either MDC1A or dystroglycanopathies, however many are being developed.

1.5.1 MDC1A

1.5.1.1 Protein therapies

Intraperitoneal or intramuscular injection of laminin-111 (α1, β1, and γ1) protein reduces muscular dystrophy and improves lifespan in the dy[W] mouse model of MDC1A.[103, 104] Although laminin-111 is not normally expressed in adult skeletal muscle,[105, 106] in this case, laminin-111 functions as a replacement for laminin-211 while reducing the possibility of triggering an immune response in laminin-211-deficient muscle. Administration of synthetic insulin-like growth factor (IGF-1) has also proven therapeutic in the dy^W mouse model.[107] For these, and all protein therapies, however, repeated administrations are required to maintain therapeutic effect.

1.5.1.2 Apoptosis inhibitors

Transgenic knock-out of the pro-apoptotic *Bax* gene,[108] and overexpression of the anti-apoptotic *Bcl2* gene[109] were therapeutic in dy^W mice. Following from the transgenic studies, administration of apoptosis inhibiting drugs omigapil[110] and doxycycline[111] also proved therapeutic in dy^W mice. Omigapil is now in clinical trials for MDC1A and collagen VI congenital muscular dystrophy.

1.5.1.3 Adeno-Associated Virus Gene therapy

Adeno-associated virus (AAV) are small, single-stranded, icosahedral, non-enveloped viruses first discovered in 1965 on electron microscopy of a simian adenovirus preparation.[112, 113] AAV was first used as a vector for transgene expression in mammalian cells in 1984.[114, 115] AAV binds various glycans at the cell surface, specific interactions being specific to AAV serotype, e.g. heparan-sulfate proteoglycans for AAVs 2, 3, and 6, and terminal galactose of N-linked glycans for AAV9.[116-119] AAV is then endocytosed and afterward released from vesicles to accumulate around the cell nucleus.[120] Wild type (wt)AAV may integrate at a specific site, AAVS1, on human chromosome 19,[121, 122] with the aid of a Rep binding element,[123] or randomly at a lower frequency. wtAAV contains a genome of about 4.7 kb with two open reading frames: the Cap gene encoding three different structural proteins via alternative splicing, and the Rep gene, encoding proteins required for replication. The wtAAV genome contains inverted terminal repeats (ITRs), self-complementary sequences that form hairpin loops at the ends of the single stranded genome, and serve as primers for second strand synthesis and mediate packaging of the viral genome within the viral capsid .[124-127] Like other viruses of the *Dependoparvoviridae*, wtAAV is replication incompetent, requiring the presence of a helper virus, such as adenovirus[114] or herpes simplex virus,[128] to complete its life cycle.

Recombinant (r)AAV are AAV in which the Rep and Cap genes have been removed and replaced with a transgene of choice (Fig. 2). They were originally created by co-transfection of the ITR-flanked transgene, Rep and Cap genes delivered in *trans* (via another plasmid), and co-infection with a helper adenovirus. This method was complicated by the need to filter out adenovirus from the preparation.[129] In this book, the triple transfection method of rAAV production was used in which the viral genome contains the therapeutic transgene, promoter and poly A elements along with the ITRS, a second plasmid is used to express the Rep and Cap genes in *trans*, and a third plasmid provides helper functions of adenovirus. [130] In gene therapy, different rAAV serotypes are used to target different tissues, e.g. AAV9 for heart, skeletal muscle, and the central nervous system,[119, 131-136] and AAV4 for lungs and kidneys.[117] Lacking the *Rep* gene, rAAV is less likely to integrate into the host genome than wtAAV, and instead the rAAV genome exists in the nucleus almost exclusively as an episome. Gene replacement therapy for disorders of small, single genes has had recent successes, for example in neuromuscular disorders like spinal muscular atrophy (SMA) and Duchenne muscular dystrophy (DMD), ocular disorders like Leber congenital amaurosis, and lysosomal storage disorders like Pompe disease.[137] Many gene therapy clinical trials have already been completed, and many more are in progress.[138]

1.5.1.4 Surrogate gene therapies

Since the *LAMA2* gene exceeds the AAV packaging capacity by almost two-fold,[139, 140] direct gene replacement is not possible, even with a dual-vector approach.[141] Surrogate gene therapy, the addition of a supporting or complementary gene instead of the mutated gene, is an alternative approach to direct replacement. Over expression of *GALGT2*, and the encoded glycosyltransferase, has shown therapeutic effect in several mouse models, including dy^W mice.[142-146] GALGT2 overexpression supports muscle by increasing expression of genes associated with developing muscle, or the NMJ, in mature extrasynaptic muscle including utrophin, agrin, and laminin-α5.[146, 147] Likewise, while laminin-α1 is normally only expressed in embryonic skeletal muscle, CRISPR-dCas9-mediated transactivation of the *Lama1* promoter in skeletal muscle caused expression of laminin-α1 and reduced pathology in a mouse model of MDC1A.[148]

Other surrogate gene therapies took advantage of the compensatory increase in laminin-α4 expression seen in laminin-α2-deficient muscle.[72] Laminin-α4 lacks the N-terminal domain required by laminin-α2 for polymerization, and has G domains with weaker binding affinity toward dystroglycan than do the G domains of laminin-α2.[149] While agrin can act as a bridge connecting laminin-α4 and dystroglycan and strengthening the total connection, it is too large to fit into the AAV vector for gene therapy. AAV-mediated overexpression of a miniaturized (mini-)agrin, however, reduced dystrophy in dy^W mice by enhancing the cumulative binding between laminin-α4 and dystroglycan.[78, 133, 150] In a different vein, expression of a chimeric "linker" protein comprised of the N-terminus of laminin-α1 and the laminin-binding N-terminus of nidogen, granted polymerization activity to the normally polymerization-incompetent laminin-α4, and reduced dystrophy in laminin-α2-deficient mouse muscle.[151, 152]

1.5.1.5 Gene correction

CRISPR-Cas9-mediated correction of *LAMA2* mutations also shows promise as a treatment for MDC1A,[153] however this strategy requires the development of unique guide RNAs for each mutation location.

1.5.2 Dystroglycanopathies

Gene replacement therapies for dystroglycanopathies are currently being developed. In some cases, gene therapy has been shown to be effective. In particular, Large overexpression can compensate for loss of other enzymes in some instances.[154, 155] Finally, as in other muscular dystrophies, corticosteroids may be used to reduce muscle inflammation and necrosis.[81, 156]

MDC1A and dystroglycanopathies have in common a pathogenic disruption of the cell membrane-ECM interface. Accordingly, I have leveraged the ECM to develop several potential gene therapies for these diseases. In this book, I conducted a series of proof of concept experiments including micro-laminin gene therapy in Chapter 2, matrix-binding trophic factors in Chapter 3, and a universal gene therapy for dystroglycanopathies in Chapter 4.

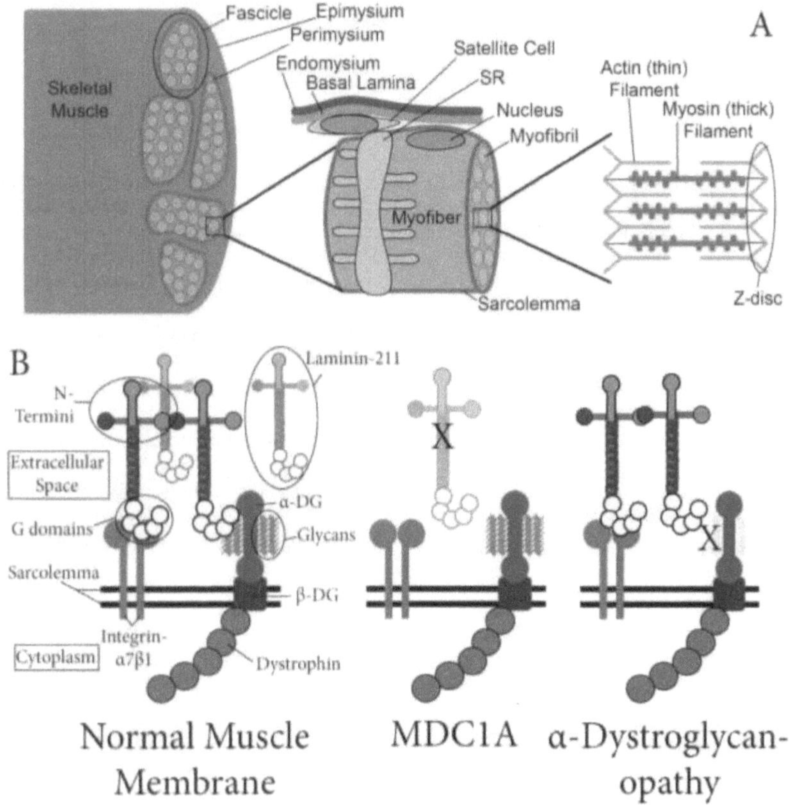

Continued.

Figure 1 Normal, MDC1A, and α-dystroglycanopathic muscle.

(A) Each muscle is surrounded by epimysium and contains perimysium-ensheathed

fascicles of myofibers. The myofibers are surrounded by endomysium, including the

Figure 1 continued.

laminin-rich basal lamina. Satellite cells are sandwiched between the myofiber

sarcolemma and the basal lamina. In healthy, developed muscle, the nuclei of myofibers

typically sit at the periphery of the cell. Myofibers are packed with myofibrils, series of

sarcomeres. The sarcomere is the contractile unit of muscle, containing actin (thin)

filaments and ATP-consuming myosin (thick) filaments. "SR" is sarcoplasmic reticulum.

(B) At the normal muscle membrane-ECM interface, laminin-211 polymerizes into a gel-

like coating by N-terminal interactions. The laminin-α2 globular (G) domains interact

with sarcolemmal receptors integrin-α7β1 and glycans attached to α-dystroglycan (α-

DG). The cytoplasmic β-dystroglycan (β-DG) C-terminus interacts with dystrophin and,

through it, the filamentous (F)-actin myofiber cytoskeleton. In MDC1A, absence of

laminin-211 disrupts the cell membrane-ECM interface. In α-dystroglycanopathy,

hypoglycosylation of α-dystroglycan disrupts laminin-211 interaction with α-

dystroglycan.

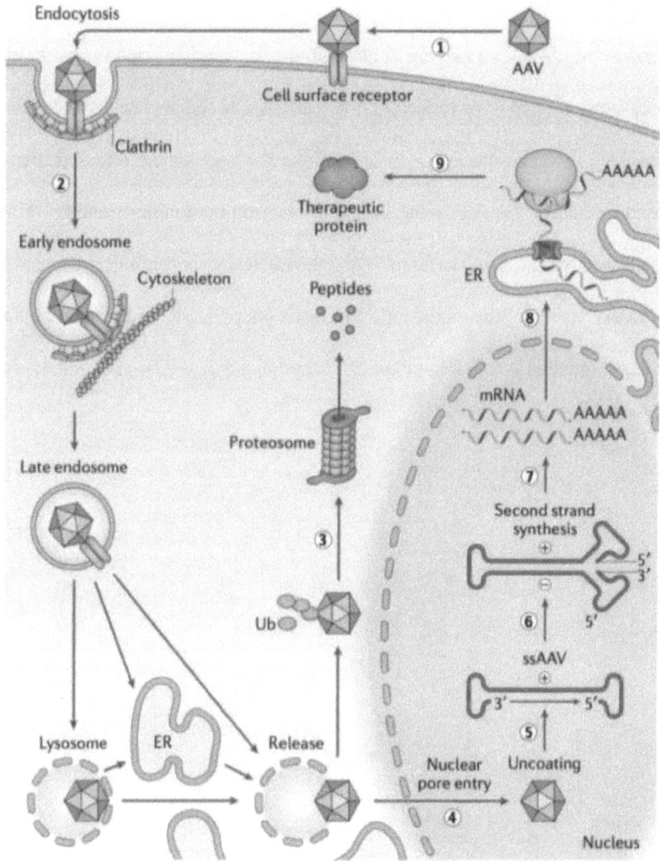

Continued.

Figure 2 rAAV-mediated expression of a therapeutic protein.

AAV vector transduction pathway. Adeno-associated virus (AAV) vector virions bind to receptors and co-receptors on the surface of target cells (step 1) and are taken into endosomes within these cells through endocytosis (step 2). Following their release from

Figure 2 continued.

endosomes, AAV virions are either ubiquitylated and targeted for proteasome- mediated

degradation (step 3) or intracellularly trafficked to the nucleus (step 4). Once in the

nucleus, AAV virions are uncoated and the AAV genome is released (step 5). The AAV

single-stranded DNA genome then is converted into double-stranded DNA (step 6),

followed by transcription (step 7) and the nuclear export of mRNA (step 8) for translation

and expression of the therapeutic transgene (step 9). ER, endoplasmic reticulum; ssAAV,

single-stranded AAV; Ub, ubiquitin. *Reproduced from Li and Samulski, 2020.*[157]

Chapter 2. Micro-laminin gene therapy can function as an inhibitor of muscle disease in the dyW mouse model of MDC1A.[1]

2.1 Introduction

Mutations in the *LAMA2* gene can cause laminin-α2-deficient Congenital Muscular Dystrophy 1A (MDC1A), [59, 60] a rare but severe disorder characterized by neonatal hypotonia, progressive muscle weakness, loss of ambulation, and premature death, often due to respiratory muscle failure. [64, 158] The *LAMA2* gene encodes laminin-α2, one subunit in the trimeric laminin-211 protein, the predominant laminin expressed in the skeletal muscle extracellular matrix (ECM). [49] Laminin-211 supports muscle structure and force transmission by anchoring myofibers to the ECM. [65, 159, 160] The N-terminal regions of the three laminin protein chains (laminin-α2, β1, and γ1) enable laminin-211 polymerization in the ECM, [52] while additional domains bind other ECM components such as agrin, [161] nidogen, and collagen IV. [56] The five globular (G) domains, unique to laminin-α subunits, bind the myofiber membrane receptors integrins [57] and dystroglycan, [162] both of which are essential to skeletal muscle membrane stability. [163, 164] Loss of laminin-211, as a consequence of *LAMA2* mutations, allows contracting myofibers to detach from the ECM and undergo apoptosis, resulting in progressive muscular dystrophy. [65] There are currently no effective FDA-approved treatments known to modify disease outcome in patients, though various strategies are being developed, including apoptosis inhibition, [108, 110, 111, 165] non-AAV gene [166] or protein [103, 104] replacement, CRISPR-mediated *LAMA2* mutation correction [153], compensatory laminin-α1 gene upregulation [167-169], and expression of mini-agrin and laminin-α1 "linker" proteins [170, 171].

At 9.5 kilobases (kb), the entire *LAMA2* gene is too large to package into a single Adeno-Associated Virus (AAV). [140, 172] Complete gene replacement is therefore impossible using AAV, unless recombination of *LAMA2* fragments is induced by expressing gene fragments in separate AAV vectors, a less efficient strategy than using a single AAV vector. [141] Gene miniaturization is one possible solution to this problem. The *DMD* gene encoding dystrophin was previously shortened into micro-dystrophin by Chamberlain and colleagues, allowing it to fit into AAV. [173] Overexpression of micro-dystrophin reduced dystrophic skeletal and cardiac muscle pathology in the mdx mouse model of Duchenne Muscular Dystrophy[173-176]. A number of other mini- or micro- dystrophin forms have been tested, some of which have shown similar therapeutic effects. [177-180] Three different AAV.micro-dystrophins are now in in phase 1/2a clinical trials (NCT03375164, NCT03368742, NCT03362502). Ruegg and colleagues have shown that expression of miniaturized agrin, termed mini-agrin, can be therapeutic in the dy^W mouse model of MDC1A. [171] This linker protein fuses the three globular (G) domains in the C-terminus of agrin, which bind muscle membrane proteins α-dystroglycan and several integrins, [181-183] to a separate region in the N-terminus of agrin that binds laminin in the ECM (the NtA domain). While this approach showed significant therapeutic effects in transgenic dy^W mice, [171, 184, 185] and also in AAV-treated dy^W mice, [133] neither strategy resulted in full recovery of normal muscle function. [171] Ruegg, Yurchenco, and colleagues have shown that co-expression of a second linker protein, αLNNd, which is a fusion of the LN domain of laminin-α1 and the laminin-binding domain of nidogen, along with mini-agrin yields additional therapeutic improvements. [170] Expression of full-length laminin-α2 or laminin-α1, in transgenic mice or via CRISPR-mediated mutation correction, also significantly inhibits disease in dy^W mice,[153, 166, 167] as does laminin-111 protein therapy. [103, 104]

Our goal here was to build on micro-gene and linker protein strategies to create an AAV-deliverable *micro-laminin* gene therapy that would allow us to test partial gene replacement in a disease model for MDC1A. In doing so, we also tested a method for increased expression of recombinant laminin-α2 G1-G5 protein fragments in the ECM. The skeletal muscle ECM contains abundant heparan sulfate glycosaminoglycans (GAGs), and these GAGs are often linked to ECM proteoglycans. [186, 187] Recently, we found that overexpression of a secreted form of the heparin-binding epidermal growth factor-like growth factor, HB-EGF, in mouse skeletal muscle led to high expression of this protein within the muscle ECM. [188] We therefore also tested whether adding the HB domain from HB-EGF to micro-laminin would increase ECM expression.

2.2 Results

2.2.1 Micro-laminins are expressed in wild type muscle after intramuscular injection of AAV

Gene therapy vectors rAAV9.CMV.*LAMA2(G1-5)* and rAAV9.CMV.*HB-LAMA2(G1-5)* expressing micro-laminins were made using the human gene segments as described in Fig. 3A. HB-LAMA2(G1-5) protein bound α-dystroglycan, a known laminin receptor,[162] in gel overlays (Fig. 3B), and both proteins displayed calcium-dependent binding in the nM concentration range to a partially purified muscle membrane preparation containing α-dystroglycan (Fig. 3C). Both HB-LAMA2(G1-5) and LAMA2(G1-5) protein were secreted from transfected CHO cells, yielding a 130 - 140 kDa protein, close to the 109 - 118 kDa weight expected from the amino acid sequence, along with proteolytic fragments on immunoblots using anti-HB-EGF and anti-human laminin-α2 G5 antibodies (Fig. 3D). HB-LAMA2(G1-5), however, also demonstrated staining of the muscle ECM in protein overlay staining experiments (Fig. 3E), which was confirmed by co-staining with laminin-α2, while LAMA2(G1-5) staining overlays did not display ECM localization when a FLAG epitope tag was on the N-terminus of this protein (Fig. 4).

To determine if micro-laminins could be expressed in the muscle ECM, we performed intramuscular (IM) injections in wild type (WT) C57Bl/6J mice. We injected 1E+11 vector genomes (vg) of rAAV9.CMV.*LAMA2(G1-5)* or rAAV.CMV.*HB-LAMA2(G1-5)* into the Tibialis Anterior (TA) muscles of 2-month-old WT mice and analyzed transduction, gene expression, and protein expression two months later. PBS was injected as a negative control, and rAAV9.CMV.*HB-EGF* (soluble form) was injected as a positive control, as we had already shown that this form of HB-EGF binds the ECM when expressed in skeletal muscle. [188] In mice injected with rAAV9.CMV.*HB-LAMA2(G1-5)*, we identified micro-laminin protein using both an anti-HB-EGF polyclonal antibody and an anti-human-specific laminin-α2 G4-5 antibody. HB-LAMA2(G1-5) staining co-localized with endogenous laminin-α2 staining in the ECM, identified with an antibody against the short-arm of mouse laminin-α2 that does not recognize the G1-G5 domains

(Fig. 5A). In mice injected with rAAV9.CMV.*LAMA2(G1-5)*, we identified micro-laminin protein by the human G4-5 domain, while no staining was evident with the HB-EGF antibody, as expected. LAMA2(G1-5) staining co-localized with mouse laminin-α2 protein in the ECM, but staining was less pronounced than for HB-LAMA2(G1-5).

We also assessed protein expression by western blotting of whole muscle protein lysates before or after heparin-agarose precipitation (Fig. 5B). rAAV9.CMV.*HB-LAMA2(G1-5)*-injected muscle produced a large amount of a 130 - 140 kDa band. This band appeared in blots using both HB-EGF and laminin-α2 G5 antibodies. The anti-laminin-α2 G5 antibody also blotted a second band of about 65 kDa, similar in size to a known ca. 80 kDa cleavage fragment previously identified in mouse muscle. [189] rAAV9.CMV.*LAMA2(G1-5)*-injected muscle also produced a 130 - 140 kDa band, reactive to the laminin-α2 G5 antibody, but not to the HB-EGF antibody. Consistent with immunostaining results (Fig. 5A), LAMA2(G1-5) band density was reduced compared to HB-LAMA2(G1-5) (Fig. 5B). Both micro-laminin proteins bound heparin-agarose in precipitation assays. There was no significant difference in vector genome (vg) muscle transduction between the rAAV9.CMV.*HB-LAMA2(G1-5)* and rAAV9.CMV.*LAMA2(G1-5)* groups (not shown). Immunoblotting for GAPDH was used as a control for equivalent protein loading and transfer.

2.2.2 Micro-laminins are expressed in dy^{W} muscle after intravenous injection of AAV

To test micro-laminin therapy in dy^{W} mice, we injected 1E+12 vg of rAAV9.CMV.*LAMA2(G1-5)* or rAAV9.CMV.*HB-LAMA2(G1-5)* intravenously (IV) into the temporal vein of postnatal day 1 (P1) dy^{W} pups. These mice were analyzed for muscle function at 2 and 3 months of age, then were euthanized at 4 months to determine transduction of tissues with AAV vector genomes (vg, Fig. 6A) as well as gene (Fig. 6B) and protein expression (Fig. 7). Although survival time is a standard outcome measurement for dy^W mice, it has been found to vary with the genetic background and, to some extent, husbandry practices between laboratories.[77] All dy^W mice in our sample groups survived up to 4 months of age, longer than expected from previous studies, and it was therefore impractical to carry out a survival study, as our other intended measurements would likely be less relevant on older tissue. Transduction of hindlimb skeletal muscles (Gastroc, TA, and Quad) was low, on the order of 0.1-0.2 vg/nucleus, as was transduction of kidney and brain (ca. 0.1 vg/nucleus). Transduction was higher in forelimb muscle (0.3-0.4 vg/nucleus, triceps brachii) and diaphragm (0.8 vg/nucleus), while transduction was highest in heart and liver (ca. 5-10 vg/nucleus). Importantly, transduction levels for mice treated with rAAV9.CMV.*HB-LAMA2(G1-5)* or rAAV9.CMV.*LAMA2(G1-5)* were roughly equivalent in most tissues. Use of the CMV promoter allowed for transgene expression throughout the body, varying in different muscles and tissues in a similar pattern to that of transduction.

Staining for micro-laminin proteins was present in the ECM surrounding the membranes of skeletal myofibers 4 months after IV delivery in P1 dy^W mice (Fig. 7A). Staining patterns in the triceps brachii muscle were similar to those observed in WT muscle following IM injection (Fig. 5A). Again, HB-LAMA2(G1-5) protein was stained by both HB-EGF and laminin-α2 G4-5 antibodies, while LAMA2(G1-5) was stained by only the laminin-α2 G4-5 antibody. As expected, residual mouse laminin-α2 protein could be detected with an anti-laminin-α2 short-arm antibody in dy^W muscle.[73] As seen in the WT

IM injections, HB-LAMA2(G1-5) staining was more intense and present in more myofibers than was LAMA(G1-5) staining (Fig. 5A).

Similar to WT IM injections (Fig. 5B), immunoblotting with a laminin-α2 G5 antibody after IV injection produced bands around 130 - 140 kDa in triceps brachii whole muscle lysate from mice receiving either vector (Fig. 7B), while immunoblotting with an HB-EGF antibody only produced a band for HB-LAMA2(G1-5). Both micro-laminin proteins could be precipitated by heparin-agarose. As before, the anti-laminin-α2 G5 antibody also recognized a ca. 65 kDa protein fragment. As with WT IM injection (Fig. 5), HB-LAMA2(G1-5) protein showed higher overall expression than LAMA2(G1-5) protein, despite no significant difference in vg transduction and transgene expression (Fig. 6). This was the case in multiple different comparisons of transduced triceps brachii muscles from rAAV9-treated dy^{W} mice (Fig. 8). The average of six such comparisons, all normalized to internal GAPDH blots, showed that overall expression of LAMA2(G1-5) protein was reduced in the triceps brachii by $35 \pm 11\%$ relative to HB-LAMA2(G1-5) (n = 5 - 6 muscles per group, $p < 0.05$).

On western blot, micro-laminin protein levels varied in muscles throughout the body (Fig. 9A). Recombinant protein was evident in the tibialis anterior (TA), quadriceps (Quad), triceps brachii (triceps), diaphragm, and heart, while expression was lower in gastrocnemius (Gastroc) muscle. Micro-laminin protein expression was not evident by western blot in liver, despite a high vg biodistribution, nor was it evident in kidney or brain (Fig. 9A). Immunostaining for HB-LAMA2(G1-5) was prominent in the diaphragm (Fig. 9B), though coincidence of HB-EGF and LAMA2 G4-5 staining was less apparent than in the triceps brachii. Staining for HB-LAMA2(G1-5) and LAMA2(G1-5) was also high in cardiomyocytes in the heart (Fig. 9C). Human laminin-α2 G4-5 staining in the liver was difficult to evaluate due to high background staining, but staining was elevated in HB-LAMA2(G1-5) relative to untreated dy^W, while LAMA2(G1-5) staining appeared to be predominantly intracellular (Fig. 9D). While endogenous laminin-α2 staining was present in the glomeruli of the WT mouse kidney, it was reduced or absent in treated and untreated dy^W mice (Fig. 9E). There was only a small degree of micro-laminin staining present in the kidney for HB-LAMA2(G1-5). Micro-laminin staining was also present in the choroid plexus of treated dy^W brain (Fig. 9F).

2.2.3 Analysis of histopathology in IV-micro-laminin-treated and control dy^W muscles

We assessed the percentage of micro-laminin-expressing myofibers and pathology within expressing triceps brachii muscles at 4 months of age after IV delivery of micro-laminin gene therapy in P1 dy^W mice. $36.3 \pm 3.8\%$ (n = 9 muscles) of myofibers stained for HB-LAMA2(G1-5) protein, while $10.6 \pm 3.2\%$ (n = 9 muscles) of myofibers stained for LAMA2(G1-5) (Fig. 10A). Within triceps brachii muscles of treated mice, myofibers expressing either micro-laminin had significantly increased diameter (Fig. 10B), reduced coefficient of variance in diameter (Fig. 10C), and reduced percentage of cells with central nuclei (a marker for a cycle of muscle degeneration and regeneration) (Fig. 10D). Increased myofiber size variability and central nucleation are indicators of muscle pathology. Thus, myofibers expressing either micro-laminin protein has less histopathology relative to non-expressing myofibers. There was, however, no significant improvement in muscle pathology between treated and untreated dy^W mice when myofibers were not broken down between expressing and non-expressing groups (Fig. 11). This may be due to the low percentage of micro-laminin-expressing myofibers or to the presence of additional regenerating fibers in treated muscles, which would be consistent with elevated overall central nuclei in HB-LAMA2(G1-5)-treated muscles.

There was no difference in muscle wasting, measured by percent muscle area occupied by myofibers, between untreated and *HB-LAMA2(G1-5)*-expressing dy^W mice, but, surprisingly, *LAMA2(G1-5)*-expressing dy^W muscle had increased muscle wasting relative to untreated muscle (Fig. 12C). To further assess fibrosis, we measured collagen-1 expression. Triceps brachii muscle in dy^W mice had regional variation in collagen-1 protein expression (Fig. 12A). In *HB-LAMA2(G1-5)*-expressing muscle, regions with greater micro-laminin protein frequently coincided with lowered collagen-1 staining, while regions with little or no HB-LAMA2(G1-5) had higher collagen-1 staining. *Col1a1* mRNA expression in the whole muscle, however, was still elevated in micro-laminin-treated dy^W muscle relative to untreated dy^W muscle and was elevated higher still in LAMA2(G1-5)-treated dy^W mice (Fig. 12B).

To assess muscle inflammation, we stained muscle for CD4, a marker for helper T cells, CD8, a marker for cytotoxic T cells, and CD68, a marker for macrophages (Fig. 13A). We then quantified the numbers of positively stained cells for each condition (Fig. 13B). There was no significant difference in the number of CD4+, CD8+, or CD68+ cells between micro-laminin-treated and untreated dy^W mice. There were, however, increases in mRNA expression for inflammatory cytokines, including interleukin β1 (*Ilb1*), C-C chemokine ligand 2 (*Ccl2*, also called MCP1), and C-C chemokine ligand 3 (*Ccl3*, also called MIP1α) in LAMA2(G1-5)-treated dy^W muscles relative to HB-LAMA2(G1-5) treatment and to untreated dy^W muscles (Fig. 13C).

We next quantified the number of intramuscular apoptotic cells using TUNEL staining with a DAPI co-stain to mark all nuclei. There was an increase TUNEL+ cells in dy^W muscle relative to WT, as previously shown,[190] and this was not decreased by either micro-laminin treatment. If anything, LAMA2(G1-5) treatment dy^W muscle trended toward higher numbers of apoptotic cells relative to HB-LAMA2(G1-5)-treated and untreated dy^W muscles (Fig. 14).

2.2.4 Improvement in forelimb grip strength in rAAV9.CMV.*HB-LAMA2(G1-5)*-treated dy^W mice, but lack of improvement in hindlimb grip strength or ambulation

We found a statistically significant improvement in body weight-normalized forelimb grip strength in dy^W mice treated with IV rAAV9.CMV.*HB-LAMA2(G1-5)* (3.1 ± 0.2 g/g, n = 12 mice) compared to untreated dy^W littermates (2.5 ± 0.1 g/g, n = 24 mice, p = 0.021) at 3 months of age (Fig. 15). Due to dystrophy and peripheral nerve demyelination,[73] dy^W hindlimbs are stiffened by paralysis and contractures. Likely because of this and lowered transduction of hindlimb muscles relative to forelimb muscles, we found that ambulation speed and duration were not significantly changed by micro-laminin treatment (Fig. 16A and B). Perhaps for similar reasons, hindlimb grip strength was not significantly increased in the HB-LAMA2(G1-5)-treated group (Fig. 16C). Untreated dy^W mice had reduced body and muscle weight compared to WT littermates. Neither micro-laminin significantly improved these measures (Fig. 17).

2.2.5 Changes in AAV biodistribution, gene and protein expression with age in micro-laminin-treated dy^W muscles

After transduction of single-stranded rAAV vectors into skeletal muscle, maximum gene expression usually takes 3 or sometimes 4 weeks. [130, 191] Therefore, we compared micro-laminin gene expression at one-month post-injection, when gene expression first reaches maximum, to expression at 4 months post-injection, the experimental endpoint, after IV injection of dy^W mice with 1E+12 vg rAAV9.CMV.*HB-LAMA2(G1-5)* at P1. Vg counts were significantly higher at 1 month of age than at 4 months of age in the TA, gastrocnemius, and triceps brachii muscles (Fig. 18A). While this reduction in vg counts over time was evident in several limb muscles, it was not observed in the diaphragm, heart, liver, or the quadriceps. Interestingly, gene expression did not necessarily follow the trend set by vg counts (Fig. 18B). Notably, the percentage of HB-LAMA2(G1-5)-stained myofibers in the triceps brachii muscle was greater at the 4-month time point (36.3 ± 3.8%, n = 9 muscles) than at the 1-month time point (11.4 ± 2.6%, n = 8 muscles, p < 0.0001) (Fig. 18C and D). Further, there was more ECM staining at 4 months than at 1 month, when staining was primarily intracellular (Fig. 18D). This result could reflect secretion of HB-LAMA2(G1-5) from transduced myofibers onto adjacent non-transduced myofibers over time (bystander effect), slow kinetics of protein expression, or increased promoter activity at the later time point. While gene expression was increased in the triceps brachii and diaphragm muscles, it was decreased in hindlimb muscles. This suggests that loss of gene expression may also occur because of ongoing muscle damage to expressing myofibers.

2.2.6 Increased expression of micro-laminin protein after IM injection in juvenile dy^W mice

We next sought to increase the percentage of myofibers expressing micro-laminin by performing IM injections in juvenile dy^W mice. To do this, we injected 1E+11 vg rAAV9.CMV.*HB-LAMA2(G1-5)* or rAAV9.CMV.*LAMA2(G1-5)* bilaterally into dy^W TA muscles at 2 weeks of age and analyzed muscles 2 months after injection. Because IM injection into a small muscle often leads to variable levels of transduction, we found vgs from rAAV9.CMV.*LAMA2(G1-5)* injection to be 2.7 ± 1-fold higher than for rAAV9.CMV.*HB-LAMA2(G1-5)*. This led to expression of micro-laminin in an equivalent percentage of myofibers for both HB-LAMA2(G1-5) and LAMA2(G1-5), even though LAMA2(G1-5) had lower expression when transduction levels were equivalent (Figs. 6, 7, and 10). In both cases, the percentage of myofibers transduced exceeded levels found with IV treatment; 45.5 ± 6.1% of muscles expressed HB-LAMA2(G1-5) protein, while 42.6 ± 6.3% of myofibers expressed LAMA2(G1-5) protein (Fig. 19A and B).

Since about half of all myofibers in the TA muscle were transduced, we assessed pathology for the entire muscle to determine the effects of each transgene on muscle pathology. Surprisingly, overall TA myofiber diameters were, on average, smaller in rAAV9.CMV.*HB-LAMA2(G1-5)* (20.8 ± 0.7 μm, n = 9 muscles, p = 0.0006) and rAAV9.CMV.*LAMA2(G1-5)* (20.6 ± 0.7, n = 10 muscles, p = 0.0003) conditions, compared to PBS-injected dy^W littermates (25.8 ± 0.8 μm, n = 12 muscles) (Fig. 19C). Within treated muscle, however, myofibers expressing *HB-LAMA2(G1-5)* or *LAMA2(G1-5)* were 54 ± 2% or 50 ± 2% larger, respectively, than non-expressing myofibers (p < 0.001

for both comparisons). Pathological measures of muscular dystrophy, including myofiber diameter variance (52.9 ± 1.0 µm/µm, n = 10 muscles, p < 0.0001) and the percentage of myofibers with central nuclei (21.5 ± 2.4%, n = 10 muscles, p = 0.0002), were reduced in muscles expressing *LAMA2(G1-5)* compared to PBS-injected dy[W] littermates (63.8 ± 1.3 µm/µm, n = 12 muscles; 25.7 ± 1.6%, n = 12 muscles) (Fig. 19D and E). Myofiber diameter variance was also reduced in muscles expressing *HB-LAMA2(G1-5)* (59.8 ± 1.2 µm/µm, n = 9 muscles). Thus, injection of muscles with AAV.micro-laminin at P14 decreased pathology measures at the whole muscle level but may have also changed postnatal muscle growth.

There are compensatory increases in other laminins, such as laminin-α4,[170] and laminin-associated proteins, in dy[W] muscles following loss of laminin-α2 expression, and upregulation of these proteins may inhibit the extent of disease. [170, 192] We therefore compared gene and protein expression for potential therapeutic molecules using TA muscles injected IM with AAV.micro-laminin (Fig. 20). Laminin-α4 (*Lama4*) gene expression was elevated about 4-fold in treated and untreated dy[W] muscles relative to WT mice, as was gene expression of *Lamb1*, which encodes laminin-ß1, while *Lama2* expression was reduced (Fig. 20B). Expression of laminin receptors, including integrin-α7ß1 (encoded by *Itga7* and *Itgb1*) and dystroglycan (encoded by *Dag1*) were not changed more than two-fold in either direction in micro-laminin-treated muscles relative to untreated dy[W] muscles. Immunostaining found differences between dy[W] TA muscles and WT TA muscle in laminin-α2, laminin-α4, integrin-α7, integrin-ß1, α-dystroglycan (by IIH6 staining), and total α/ß-dystroglycan expression. Comparison of western blots for these proteins between treated and untreated TA muscles, using GAPDH as an internal control for protein loading and transfer (which was lower in dy^W than in WT but which did not significantly vary between treated and untreated dy^W muscles), showed that laminin-α4 expression was elevated, though not to a statistically significant level, in *HB-LAMA2(G1-5)-* and *LAMA2(G1-5)-*treated dy^W muscle relative to untreated dy^W muscle, as was expression of α-dystroglycan (by IIH6 blotting) and ß-dystroglycan (Fig. 20C - F). Expression of integrin-α7 and -ß1 was more variable. We used a polyclonal antibody to the C-terminus of integrin-α7 for immunoblotting that recognizes multiple cleaved protein forms in addition to the native 120kDa protein, resulting in multiple bands, and so we quantified each protein fragment separately.

2.3 Discussion

While *LAMA2* whole gene replacement is known to be therapeutically efficacious in mouse models of MDC1A using transgenic or gene editing approaches, [153, 166] the *LAMA2* gene is too large to fit into a single AAV vector, the best current clinical method for gene therapy. In this work, we have designed a micro-laminin gene therapy and demonstrated that it can prevent several aspects of muscle histopathology and even in one instance improve muscle function in the dy^W mouse model for MDC1A. This strategy is akin to the linker protein therapies used by Ruegg, Yurchenco and colleagues to create a *LAMA2* surrogate out of forms of agrin [171] or laminin-α1. [170] In both cases, functional protein-binding regions were linked to an ECM-binding region to create structural scaffoldings akin to those present in the native laminin-211 protein. We have used a similar linker protein strategy with the G1-G5 domains of *LAMA2* protein itself, which allows for a partial gene replacement, by adding the heparin-binding (HB) motif from HB-EGF, a protein that can be highly expressed in the muscle ECM. [188] Interaction of the HB domain with muscle ECM likely results from binding to GAGs that are copiously present in the muscle ECM. [187] In doing so, this very small HB domain may replace the much larger ECM-binding domains in the N-termini of the $\alpha2$, $\beta1$, and $\gamma1$ chains of laminin-211 that normally dictate ECM localization. [52, 56, 193] The G1-G5 region of *LAMA2* also has the ability to bind heparin,[194, 195] , but this heparin binding region likely overlaps with its dystroglycan binding domain. [195, 196], which can be blocked by heparin. Thus, addition of the HB domain may free the G1-5 domains of laminin-$\alpha2$ to properly bind dystroglycan at the muscle membrane. The HB domain may also serve to facilitate the secretion of the large G1-G-5 protein fragment, which is less well expressed and less effective, in some instances even deleterious, in muscle in the absence of the HB domain.

While our gene therapy approach is the first to show that the micro-laminin can have clinical benefits in a mouse model for MDC1A, it clearly is only one step toward maximizing this approach for patients. For example, while micro-laminin expression could prevent muscle damage, increase myofiber size, and even increase weight-normalized forelimb grip strength, dy^W mice treated with micro-laminin displayed no improvement in ambulation or hindlimb grip strength. This may be due to limited treatment effect on the peripheral nervous system, as dy^W mice display demyelination of peripheral nerves and gradual hindlimb paralysis, [197] or it may be due to the limited transduction in hindlimb muscles. Additionally, the AAV vectors we have used may need to be optimized further to allow more efficient protein expression and secretion. The expression of such a large ECM protein clearly requires time. Analysis of HB-LAMA2(G1-5) staining at 1 month post-treatment, when gene expression has become saturating in dy[W] mice, [191] showed a protein expression pattern far less coincident with the ECM, while staining at 4 months post-treatment showed higher ECM localization. dy[W] mice have reduced ability to regenerate muscle and display increased muscle cell apoptosis, both of which contribute to increased muscle histopathology. [108, 165, 198-200] The kinetic lag in achieving maximum gene expression[191] and full protein expression using AAV may have worked against our testing of this therapy in the dy[W] model. Immune responses to the human laminin protein, or HB-LAMA2 fusion domain epitopes, may also have worked against achieving maximal protein expression.

In its current form, micro-laminin gene therapy does not compare favorably to other treatment strategies tested in dy[W] mice including rAAV9-delivered mini-agrin,[133] apoptosis inhibitor omigapil,[110, 201] and transgenic expression of the linker protein αLNNd. Each of these treatments achieved improvements in body weight, grip strength, and/or ambulation and reduces fibrosis and/or muscle cell apoptosis. By contrast, HB-LAMA2(G1-5) micro-laminin gene therapy demonstrated limited therapeutic effect in dy[W] mice, having no effect on muscle cell apoptosis or muscle inflammation, but it was successfully expressed, localized to muscle ECM, reduced some aspects of muscle histopathology. It also was generally more effective than LAMA2(G1-5) without the HB domain, suggesting that additional modifications to improve protein secretion, expression, and functionality may further improve treatment outcomes for MDC1A.

2.4 Materials and Methods

2.4.1 Production of AAV9 vectors containing micro-laminin genes

The LAMA2(G1-5) and HB-LAMA2(G1-5) micro-laminin genes were designed using the human LAMA2 G domain coding sequence from RefSeq NM_000426.3 (base pairs 6538 to 9436), the human HBEGF signal peptide from NM_001945.2 (base pairs 276 to 348) for LAMA2(G1-5), and the human HBEGF signal peptide through heparin-binding domain from NM_001945.2 (base pairs 276 to 597) for HB-LAMA2(G1-5). The recombinant genes were synthesized by GeneArt Gene Synthesis (Regensburg, Germany) in a pMA-T vector backbone. The genes were subcloned into a pcDNA3.3-TOPO backbone for use in cell culture transfections, and into a rAAV plasmid with CMV promoter, SV40 enhancer, and SV40 poly-adenylation signal for vector production. From 5' ITR to 3' ITR, the length the genome length of rAAV.CMV.HB-LAMA2(G1-5) was 4.649 kb, and the length of rAAV.CMV.LAMA2(G1-5) was 4.377 kb. Plasmids were packaged in rAAV9 vectors by the Nationwide Children's Hospital Viral Vector Core (Columbus, OH) using the triple transfection technique in HEK293 cells,[130] and highly purified using sucrose density centrifugation and anion exchange chromatography, as previously described.[202]

2.4.2 Cell Transfection

CHO-K1 cells (ATCC, Manassas, Virginia) were grown in 10 cm dishes in DMEM (Thermo Fisher Scientific) with 10% fetal bovine serum (Millipore-Sigma). When at 80% confluence, cells were transfected with pcDNA3.3-TOPO.CMV.*HB-LAMA2(G1-5)* or pcDNA3.3-TOPO.CMV.*FLAG-LAMA2(G1-5)* using *Trans*IT-X2 Dynamic Delivery System (Mirus, Madison, Wisconsin), according to the manufacturer's instructions for CHO-K1 cells in 10 cm dishes.

2.4.3 Mice

All mouse experiments were performed using protocols approved by the Institutional Animal Care and Use Committee at The Abigail Wexner Research Institute at Nationwide Children's Hospital (Columbus, OH). C57BL/6J mice were purchased from Jackson Laboratories (Bar Harbor, ME). dy^W mice bred on a pure C57BL/6 background were a generous gift from Dr. Dean Burkin at University of Nevada, Reno, Medical School.

2.4.4 Intramuscular AAV Injections

Adult mice were anesthetized with isofluorane (Baxter, Deerfield, IL), and hindlimbs were shaved over the TA muscles. 1E+11 vg rAAV9 in 25 μL PBS was injected into each TA muscle using a 0.33cc Thinpro Insulin Syringe (Terumo Medical, Somerset, NJ). Two-week-old mice were restrained and 1E+11 vg rAAV9 in 5 μL PBS was injected into each TA muscle using a 0.3cc Thinpro Insulin Syringe.

2.4.5 Intravenous AAV Injections

Mouse pups up to 48 hours old were anesthetized on ice and injected with 1E+12 vg rAAV9 into the temporal vein using a Thinpro 0.3cc Insulin Syringe. The minimum injection volume allowable by virus titer was used.

2.4.6 Grip Strength

On three consecutive days, one week prior to grip strength assessment, mice were trained on the Grip Strength Meter (Columbus Instruments, Columbus, OH). Each training day consisted of five forelimb and five hindlimb grip strength trials which were not recorded. The week of assessment, grip strength was assessed on five consecutive days. Each assessment consisted of five forelimb trials using the "T Peak" setting and five hindlimb trials using the "C Peak" setting on the Grip Strength Meter and measured in grams, followed by body weight measurement. For each mouse, on each day of the assessment week, the five trials were averaged and normalized to body weight obtained on that day. For each mouse, the final grip strength value was the average of the body weight-normalized grip strength values obtained over the five days.

2.4.7 Ambulation

On three consecutive days, one week prior to ambulation assessment, mice were trained on the Treadmill Meter (Columbus Instruments, Columbus, OH). The week of assessment, mice were placed on the stationary Treadmill Meter, one or two mice per enclosed lane. The treadmill was started at 5 meters per minute, increasing 1 meter per minute every 30 seconds until 10 meters per minute was reached. The treadmill was run at 10 meters per minute for 10 minutes. For each mouse, the trial ended when the mouse fell off the end of the treadmill three times in a span of 20 seconds. At the end of the trial, the maximum speed reached and time of ambulation at 10 meters per minute were recorded. If a mouse failed to reach 10 meters per minute, 0 seconds was recorded for ambulation time. For each mouse, the final values for speed and ambulation time were the averages of those obtained over the five days.

2.4.8 Tissue Processing

Skeletal muscles were removed, weighed, mounted with O.C.T. (Thermo Fisher Scientific) on cork, and snap-frozen liquid nitrogen-cooled 2-methylbutane. The internal organs and heart were removed, placed in Peel-A-Way Disposable Embedding Molds (Thermo Fisher Scientific), covered in Tissue Plus O.C.T. (Thermo Fisher Scientific) and frozen in a 2-methylbutane (Thermo Fisher Scientific) / dry ice bath. One hemisphere of the brain was snap-frozen for molecular analysis. The other hemisphere was fixed overnight in 4% paraformaldehyde (EMS, Hatfield, PA) in PBS at 4°C, washed overnight 0.1 M glycine (Thermo Fisher Scientific) in PBS at 4°C, then incubated sequentially in 10%, 20%, and 30% sucrose each overnight at 4°C, before freezing in an embedding mold covered with O.C.T.

2.4.9 AAV Biodistribution

DNA was extracted from tissue shavings using a DNeasy Blood & Tissue Kit (Qiagen, Germantown, MD). Vector genome quantity in 100 ng of sample DNA was determined using 2X TaqMan Gene Expression Master Mix, a standard curve of linearized rAAV plasmid from 5 vg to 5E+6 vg, and the following primer/probe set amplifying a region spanning the SV40 enhancer and the transgene: 5'-ACTTCTAGGCCTGTACGGAAGTG-3' (forward), 5'- 56-FAM/AAAGCTGCG/ZEN/GAATTGTACCCGCGGC/3IABkFQ/-3' (probe), and 5'-CTGCAGCCAGAAAGAGCTTCAG-3' (reverse). Vector genomes per mouse nucleus were then calculated using an estimate of 1.7E+5 mouse nuclei per 1 µg of genomic DNA. **2.4.10 Gene Expression Measurement**

RNA was extracted from tissue shavings using a Direct-zol RNA MiniPrep kit (Zymo Research, Irvine, CA), and cDNA was prepared using a High-Capacity cDNA Reverse Transcriptase kit (Applied Biosystems, Bedford, MA). *HB-LAMA2(G1-5)* and *LAMA2(G1-5)* transgene expression was measured in cDNA samples using TaqMan Gene Expression Master Mix and *LAMA2* (human) Assay Hs01124072_m1 (Applied Biosystems). Additional assays included mouse *Lama4*, mouse *Itga7*, and mouse *Itgb1* (IDT, Coralville, IA), mouse *Lamb1* Assay Mm008801853_m1, and mouse *Dag1* Assay Mm00802400 (Applied Biosystems). 18S rRNA was used as an internal control (Applied Biosystems). The $2^{-\Delta\Delta Ct}$ method was used to calculate fold-change in expression. [203]

2.4.11 WGA-Agarose Precipitation and Protein Isolation

Immediately following dissection, gastrocnemius and quadriceps muscles from WT mice were minced with a sterile scalpel, placed in 5 mL NP40 lysis buffer, 1% (w/v) NP40 (Millipore-Sigma, St. Louis, MO) in 50 mM Tris buffer with 150 mM NaCl, 1 mM EDTA (Invitrogen, Grand Island, NY), and cOmplete EDTA-free Protease Inhibitor Cocktail (Millipore-Sigma), and rocked overnight at 4°C. Supernatant was transferred to 200 µL WGA-agarose (EY Laboratories, San Mateo, CA), and rocked overnight at 4°C. Precipitated sample was washed thoroughly in 0.1% (w/v) NP40 in 50 mM Tris buffer with 150 mM NaCl, then eluted in 500 µL 0.3 M *N*-Acetyl-D-Glucosamine (GlcNAc, Millipore-Sigma) in 0.1% NP40 buffer for 20 minutes with rocking at 4°C. To reduce GlcNAc concentration, eluted samples were dialyzed in 1000-fold volume 0.1% NP40 buffer overnight at 4°C using 10K MWCO Slide-A-Lyzer Dialysis Cassettes (Thermo Fisher Scientific).

2.4.12 Heparin-Agarose Precipitation and Protein Isolation

200 µL Heparin-Agarose Type I, saline suspension (Millipore-Sigma) was added per mL of conditioned media, cell lysate, or muscle lysate to be precipitated, and mixed overnight at 4°C. Precipitated samples were washed thoroughly with PBS. For western blotting, precipitated samples were resuspended in 50 µL 1X NuPage LDS Sample Buffer with 100 mM β-mercaptoethanol. 20 µL of the resuspension was boiled and used for immunoblotting. For isolation of micro-laminin proteins, precipitated samples were eluted with a high salt solution, 200 µL 1.5 M NaCl in 0.01 M Tris-HCl, pH 7.5. To reduce salt concentration, eluted samples were dialyzed in 1000-fold volume 0.01 M Tris-HCl overnight at 4°C using 10K MWCO Slide-A-Lyzer Dialysis Cassettes (Thermo Fisher Scientific).

2.4.13 Anti-FLAG-Agarose Precipitation and Protein Isolation

Performed identically to heparin-agarose precipitation and protein isolation, replacing Heparin-Agarose Type I, with Anti-FLAG M2 Affinity Gel (Millipore-Sigma). High salt solution was used for elution, as in heparin-agarose precipitation and isolation.

2.4.14 Immunoblotting

Protein was extracted from tissue shavings using NP40 lysis buffer. 40 µg of each protein lysate was added to NuPAGE LDS Sample Buffer (Thermo Fisher Scientific) with 100 mM β-mercaptoethanol (Thermo Fisher Scientific) and boiled for 10 minutes prior to gel electrophoresis and transfer to nitrocellulose membranes for immunoblotting. The antibodies against HB-EGF (AF-259), laminin-α4 (AF3837), and α-/β-dystroglycan (AF6868) were obtained from R&D Systems (Minneapolis, MN). The antibody against laminin-α2 G5 (2D4) was obtained from Abnova (Taipei, Taiwan). The antibodies against GAPDH (G9545) and α-dystroglycan, clone IIH6C4 (05-593) were obtained from Millipore-Sigma. The antibody against laminin-β1 (PA5-27271) was obtained from Invitrogen. The antibodies against integrins-α7B and -β1 were obtained as described in Martin et. al. [204] The HRP-conjugated secondary antibodies against goat and rabbit IgG were obtained from Jackson ImmunoResearch (West Grove, PA), and the secondary antibody against mouse IgG1 was obtained from Bethyl Laboratories (Montgomery, TX). Protein band densities were measured using ImageJ 1.46r.

For protein overlays, protein gels were transferred to Immobilon-PSQ PVDF membranes (Millipore-Sigma) and washed briefly in laminin-binding buffer (LBB), 140 mM NaCl, 1mM CaCl$_2$-, and 1mM MgCl$_2$ in 10 mM Tris-HCl, before blocking in 5% milk in LBB. Membranes were overlaid with micro-laminin protein samples in LBB overnight at 4°C. Primary antibody in 3% BSA in LBB was added for 2 hours at room temperature, followed by secondary antibody in 3% BSA in LBB for 1 hour.

2.4.15 Enzyme-Linked Immunosorbant Assays (ELISAs)

200 ng WGA-agarose-isolated sample was diluted in 50 mM bicarbonate buffer, pH 9.5 and added per-well to NUNC-Immuno MicroWell plates, then incubated overnight at 4°C. Plates were blocked in 3% BSA in 50 mM Tris-HCl with 100 mM NaCl, 1mM CaCl$_2$, and 1 mM MgCl$_2$, then incubated with micro-laminin ligands in blocking buffer. Ligands were detected using anti-laminin-α2 G5 (2D4) primary antibody and HRP-conjugated secondary antibody. Substrate Reagent Pack colorimetric kit (R&D Systems) was used to measure signal at 450 nm.

2.4.16 Immunofluorescent Staining

10 μm tissue sections, cut on a Cryostat, were placed on Fisher Superfrost Plus glass slides (Thermo Fisher Scientific). A hydrophobic border was drawn around each section using an ImmEdge Hydrophobic Barrier PAP pen (Vector Labs, Burlingame, CA). Sections were fixed in 2% Paraformaldehyde (EMS) in PBS for 10 minutes, permeabilized in 0.1% Triton X-100 (Millipore-Sigma) in PBS for 10 minutes, incubated in Mouse-on-Mouse Blocking Reagent (Vector Labs) for 1 hour, and blocked with 5% donkey serum in PBS for 30 minutes prior to incubation overnight at 4°C in primary antibodies. Sections were then incubated with secondary antibodies for 1 hour at room temperature and then treated with ProLong Gold with DAPI (Invitrogen) before adding a coverslip. The antibodies against HB-EGF (AF-259), laminin-α4 (AF3837), integrin-α7 (MAB3518), and α-/β-dystroglycan (AF6868) were obtained from R&D Systems. The antibodies against human laminin-α2 G4-5 (5H2), laminin-α2 short-arm (4H8-2), and α-dystroglycan, clone IIH6C4 (05-593) were obtained from Millipore-Sigma. The antibody against dystrophin (ab15277) was obtained from Abcam (Cambridge, UK). The antibody against integrin-β1 clone 9EG7 (553715) was obtained from BD Biosciences (San Jose, CA). The antibody against laminin-α5 (bs-1086R) was obtained from Bioss (Boston, MA). Rabbit anti-mouse IgG1 secondary antibody [FITC] (NBP1-73636) was obtained from Novus Biologicals (Centennial, CO). All other secondary antibodies were obtained from Jackson ImmunoResearch.

For protein overlay assays, following blocking, sections were incubated in heparin-agarose-isolated micro-laminin protein from transfected cell lysate in 3% BSA in LBB overnight at 4°C. Primary antibody in 3% BSA in LBB was added for 2 hours at room temperature, followed by secondary antibody in 3% BSA in LBB for 1 hour.

2.4.17 Microscopy and Imaging

For all immunofluorescent images, a Zeiss Imager.Z1 with an AxioCam MRm camera, Plan-Apochromat 10x/0.45 M27 and Plan-Apochromat 20x/0.8 M27 objectives, and AxioVision40 v4.7.1.0 software was used to take single-plane multi-channel epifluorescence images. Equal exposure times were used to image corresponding channels across samples within each experiment. In all staining, DAPI and Alexa 488, CY2, or FITC, CY3, and CY5 fluorophores were used and visualized using fluorophore-appropriate filters.

2.4.18 Histological Analysis

Myofiber diameter, central nucleation, and muscle wasting analyses were performed using an ImageJ macro on 10X images of muscle sections stained with an antibody against dystrophin to outline myofibers, much as previously described. [205] TUNEL(+) nuclei were counted using an ImageJ macro on 10X images of muscle sections. Immune cells were counted manually on 10X images of muscle sections.

2.4.19 Statistical Analysis

The statistical significance of differences between two groups was tested using an unpaired, two-tailed Student's t-test. The significance of differences between more than two groups was tested using a one-way ANOVA, with post-hoc Tukey test for multiple comparisons.

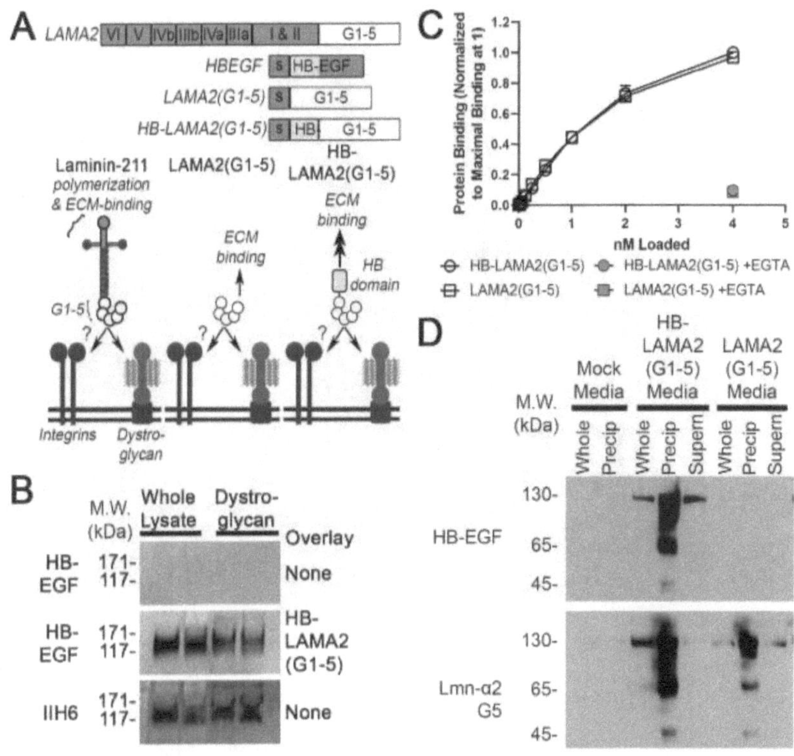

Continued.

Figure 3 Micro-laminin design and in vitro characterization

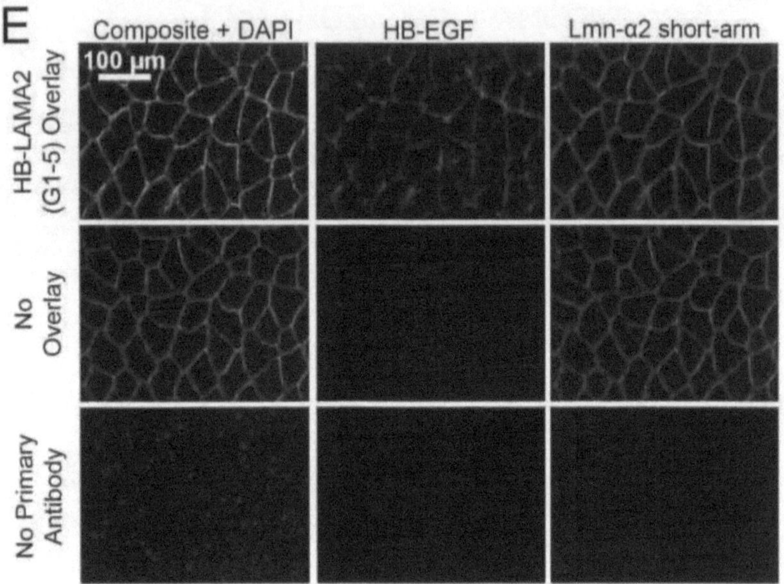

Continued.

(A) Using human gene segments, *LAMA2(G1-5)* was designed by adding the *HBEGF* gene segment encoding the signal peptide (s) to the 5' end of the *LAMA2* gene segment encoding G domains 1-5. *HB-LAMA2(G1-5)* was designed by adding the *HBEGF* gene segment encoding the signal peptide, prepropeptide domain (which is normally cleaved from the protein), and the heparin binding (HB) domain to the 5' end of the *LAMA2* gene segment encoding G domains 1-5. The HB domain of HB-EGF was expected to enhance laminin-α2 G1-5 binding to the muscle ECM. The G1-G5 domains of LAMA2 were expected to bind muscle membrane receptors including integrins and dystroglycan. (B) Whole muscle

Figure 3 continued.

lysates and WGA-purified dystroglycan were overlaid with HB-LAMA2(G1-5) protein, or incubated in dilution buffer alone (None), then immunoblotted with antibody recognizing the HB-domain of HB-EGF. IIH6 immunoblot recognized α-dystroglycan and demonstrated similar band location to HB-LAMA2(G1-5) overlay. (C) ELISA binding assay of HB-LAMA2(G1-5) and LAMA2(G1-5) proteins to WGA-purified muscle membrane proteins including α-dystroglycan. Protein binding in buffer containing 1 mM Ca^{2+} and 1 mM Mg^{2+} (open circles and squares) was compared to binding in buffer with 10 mM EGTA (filled circles and squares). (D) Whole conditioned media (Whole), media precipitated by heparin-agarose (Precip), or supernatant following precipitation (Supern) from CHO cells transfected with *HB-LAMA2(G1-5)* or *LAMA2(G1-5)* expression vectors was compared to media from mock-transfected controls. Secreted protein was recognized by blotting with anti-HB-EGF or anti-human laminin-α2 G5 (Lmn-α2 G5) antibodies. From amino acid sequence, the expected weight of HB-LAMA2(G1-5) is 118 kDa and that of LAMA2(G1-5) is 109 kDa. (E) Muscle sections were overlaid with HB-LAMA2(G1-5) protein and protein binding was visualized by immunostaining with an anti-HB-EGF antibody, then compared to staining for endogenous laminin-α2 using an antibody to the short-arm domain not present in micro-laminin. In B, D, and E, antibodies used for staining or blotting are shown on the left side of the figures

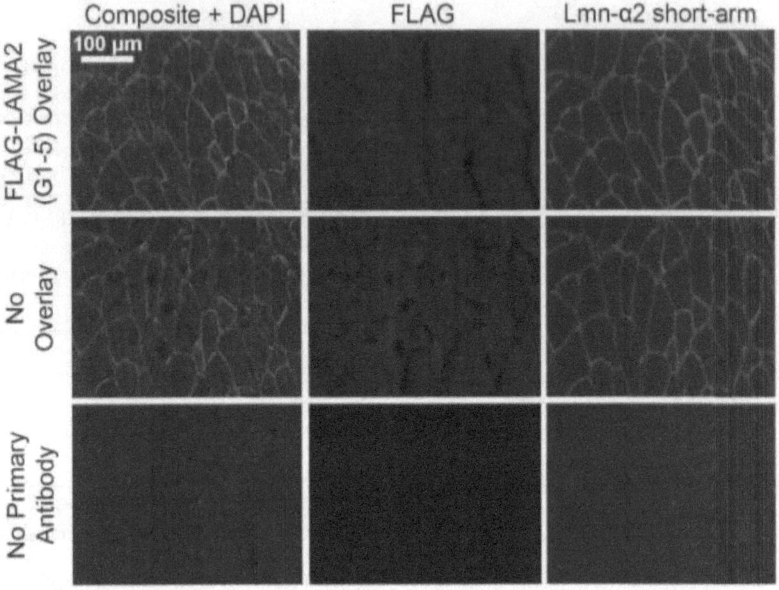

Figure 4 LAMA2(G1-5) protein staining of muscle sections

Muscle sections were overlaid with FLAG-LAMA2(G1-5) protein, and protein binding

was visualized by immunostaining with an anti-FLAG antibody, then compared to staining

for the laminin-α2 short-arm not present in micro-laminins.

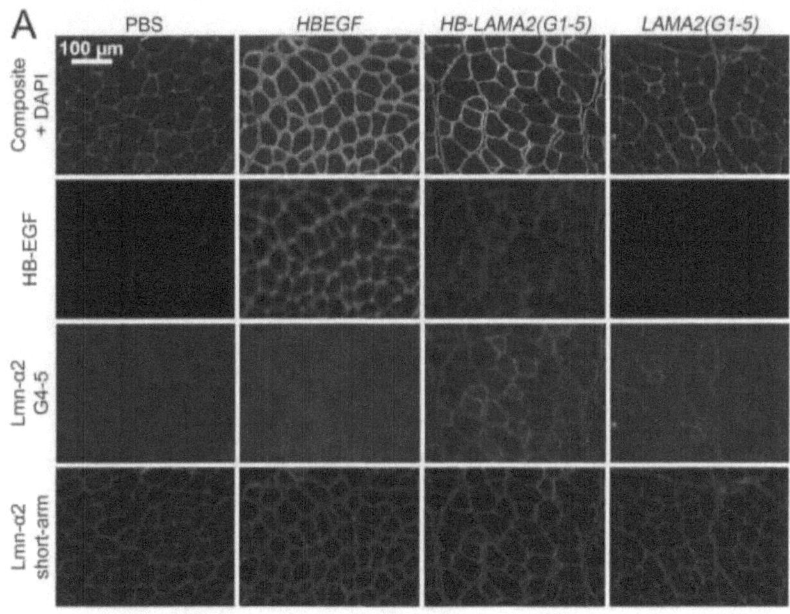

Continued.

Figure 5 Micro-laminin protein expression in mouse skeletal muscle following intramuscular injection

Figure 5 continued.

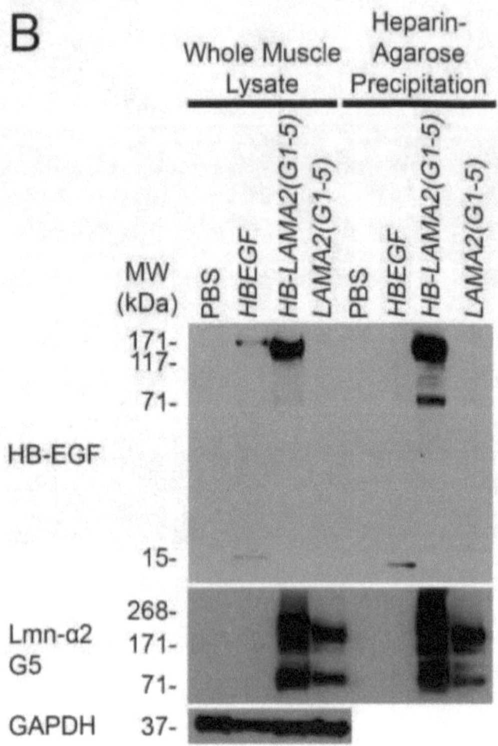

Continued.

TA muscles from adult wild type mice were injected with 1E+11 vg rAAV9.CMV.*HBEGF*, rAAV9.CMV.*HB-LAMA2(G1-5)*, rAAV9.CMV.*LAMA2(G1-5)*, or an equivalent volume of PBS. Muscles were analyzed at 2 months post-injection. (A) TA muscle sections stained with antibodies against HB-EGF (red), human laminin-α2 G4-5 (Lmn-α2 G4-5, green), and mouse laminin-α2 short-arm (Lmn-α2 short-arm, far-red).

Figure 5 continued.

(B) Recombinant human HB-EGF protein (rHB-EGF) and TA muscle lysates were blotted with antibodies against HB-EGF and human laminin-α2 G5 domain (Lmn-α2 G5). Anti-GAPDH blotting was performed as a loading control. Heparin-agarose precipitation of muscle lysates was performed to demonstrate heparin-binding. Genes expressed are shown on the top of the figures, with PBS injection shown as a negative control. Antibodies used for staining or blotting are shown on the left side of the figures.

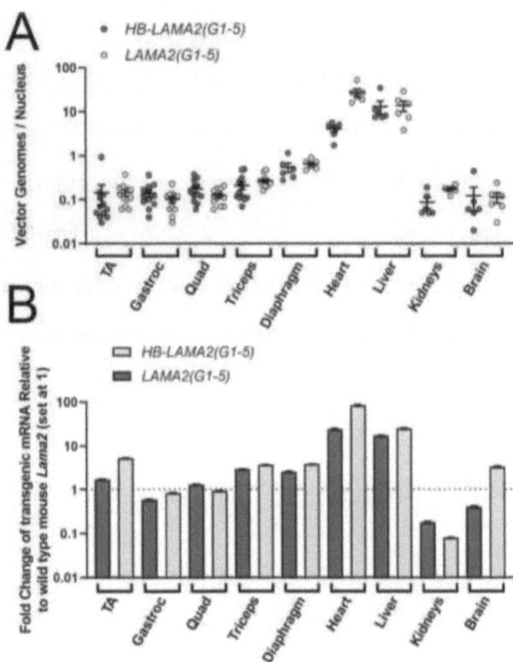

Figure 6 Biodistribution of AAV vector genomes and transgene expression in dy^W mice following intravenous injection

Neonatal dy^W mice were injected intravenously with 1E+12 vg rAAV9 containing the micro-laminin genes under the CMV promoter. DNA was extracted from tissues at 4 months of age. (A) The number of AAV vector genomes per nucleus was quantified using qPCR. Errors bars are standard error of the mean (SEM) for n = 6 mice per condition. (B) Fold change in transgene expression was calculated relative to expression of the endogenous mouse *Lama2* gene in corresponding tissues of age-matched WT littermates, with 18S ribosomal RNA gene expression as an internal control. Errors bars are SEM for n = 6 mice per condition.

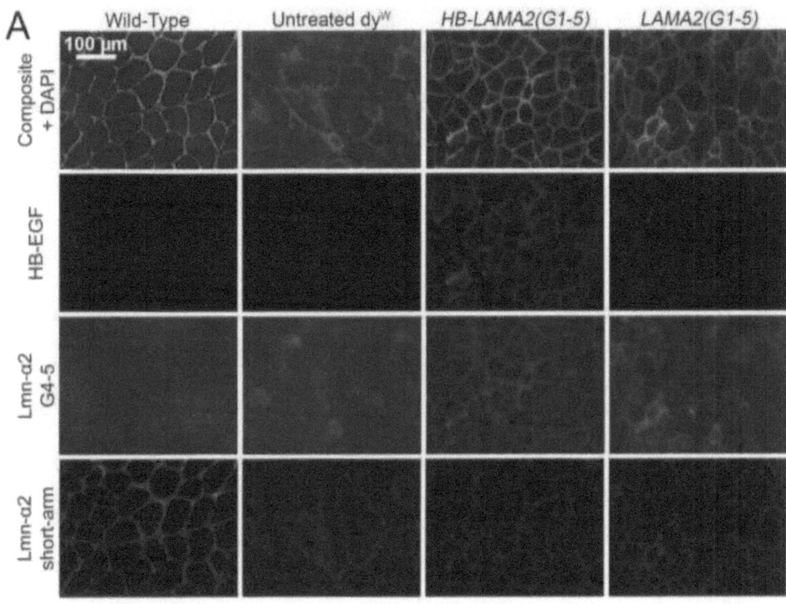

Continued.

Figure 7 Micro-laminin protein expression in dy^W mouse skeletal muscle following intravenous AAV injection

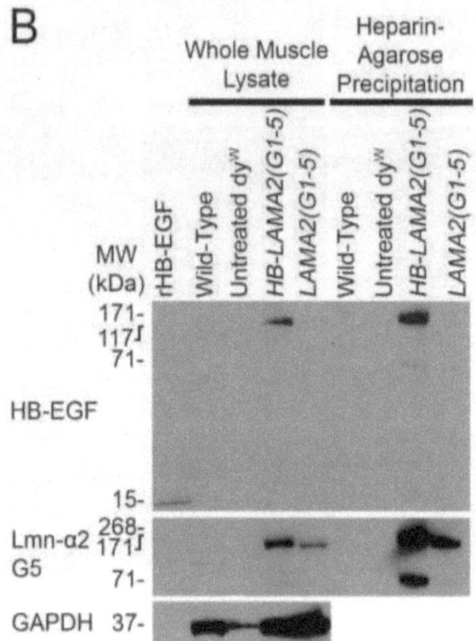

Tissues from dy^W mice treated with intravenous rAAV9, and from untreated WT and dy^W littermates, were collected at 4 months of age. (A) Triceps brachii muscle sections stained with antibodies against HB-EGF (red), human laminin-α2 G4-5 (Lmn-α2 G4-5, green), and mouse laminin-α2 short-arm (Lmn-α2 short-arm, far-red). (B) Recombinant human HB-EGF protein and triceps brachii muscle lysates were immunoblotted with antibodies against HB-EGF and human laminin-α2 G5. Anti-GAPDH blotting was performed as a loading control. A heparin-agarose precipitation of triceps brachii muscle lysates was performed to demonstrate binding to heparin. Antibodies used for staining or blotting are shown on the left side of the figures.

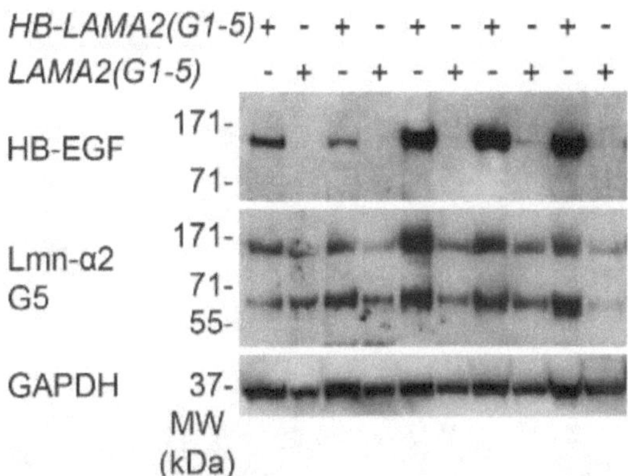

Figure 8 Micro-laminin protein expression in dy^W mice treated with rAAV9.CMV.*HB-LAMA2(G1-5)* or rAAV9.CMV.*LAMA2(G1-5)*

Triceps brachii muscle lysates from five dy^W mice treated intravenously at P1 with rAAV9.CMV.*HB-LAMA2(G1-5)* or rAAV9.CMV.*LAMA2(G1-5)* were compared by immunoblotting at 4 months of age. GAPDH was blotted as a control for protein loading and transfer.

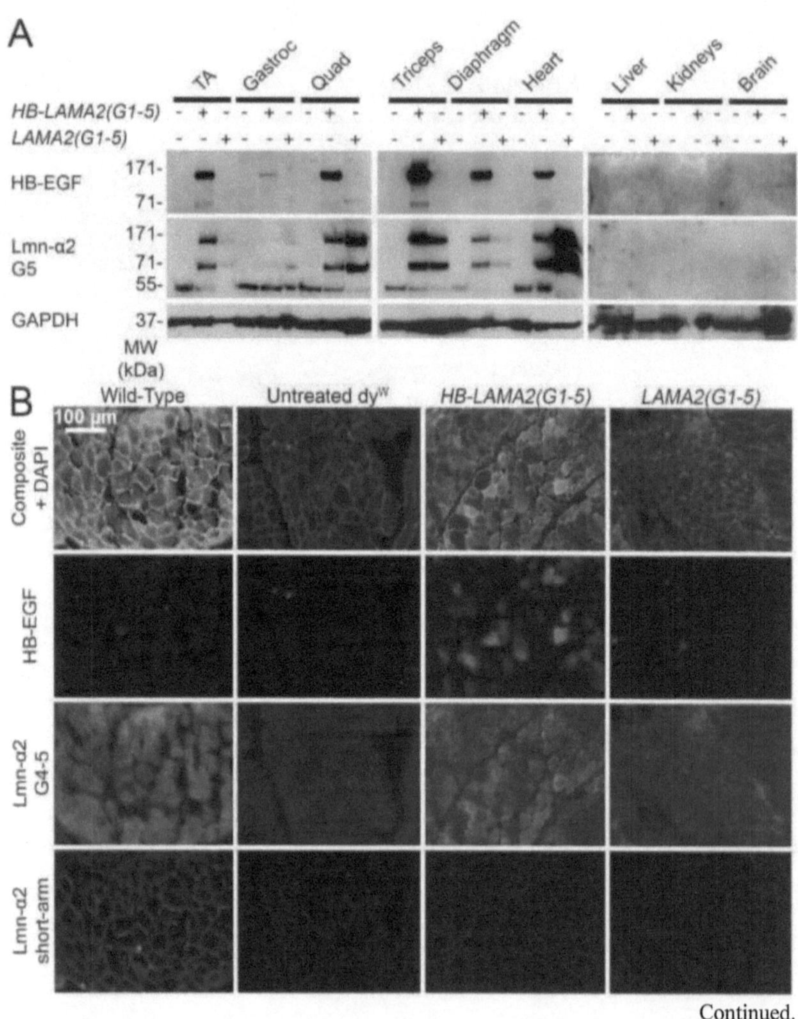

Continued.

Figure 9 Micro-laminin protein expression throughout dy^W mouse tissues following intravenous rAAV9 injection

Figure 9 continued.

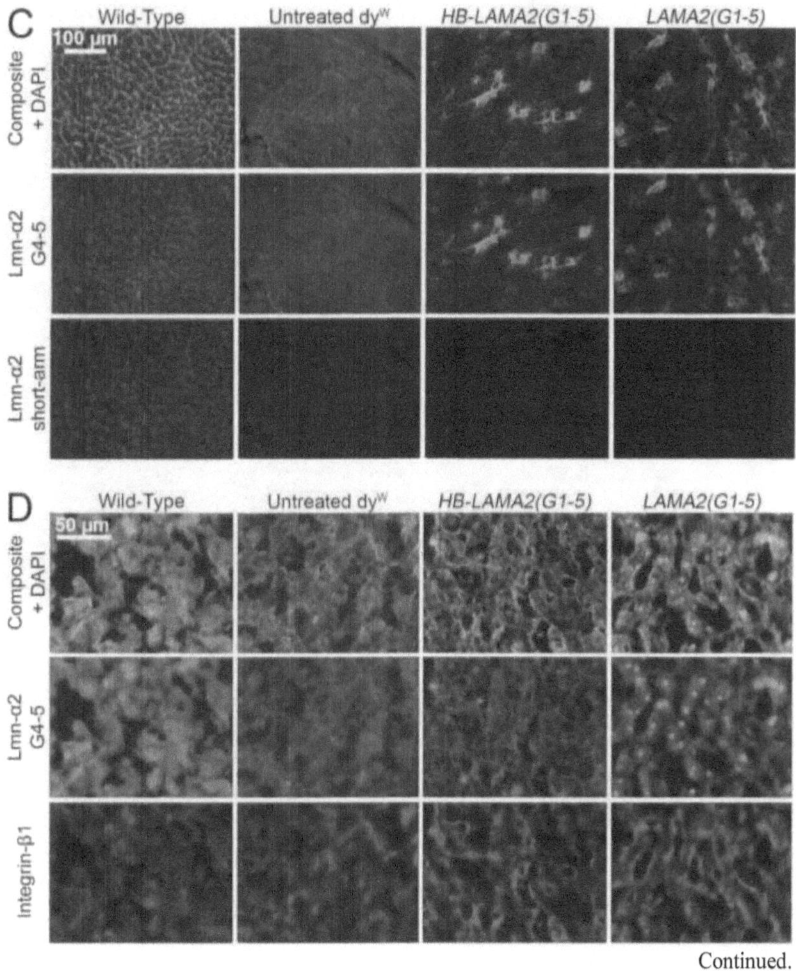

Continued.

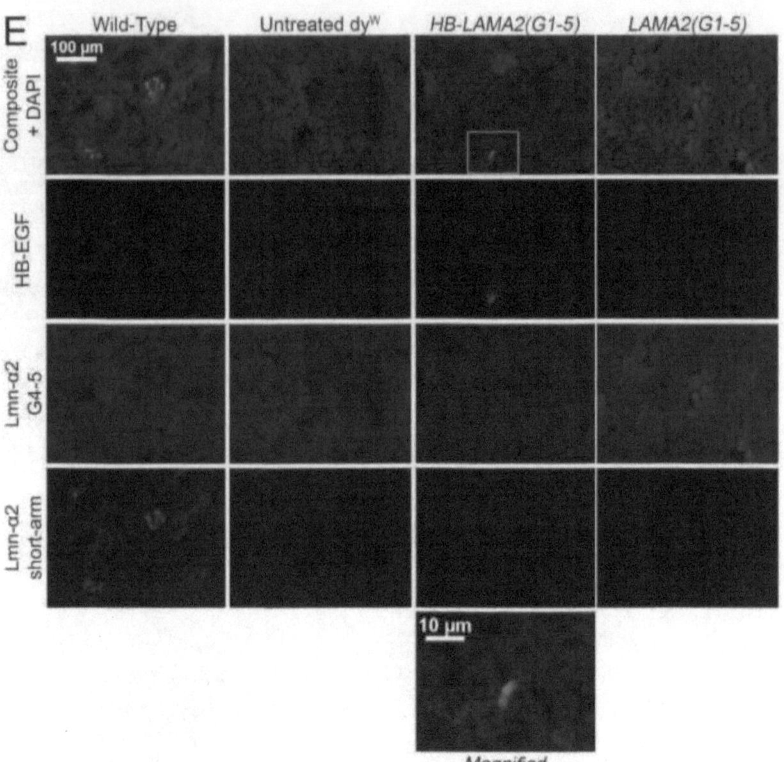

Continued.

Figure 9 continued.

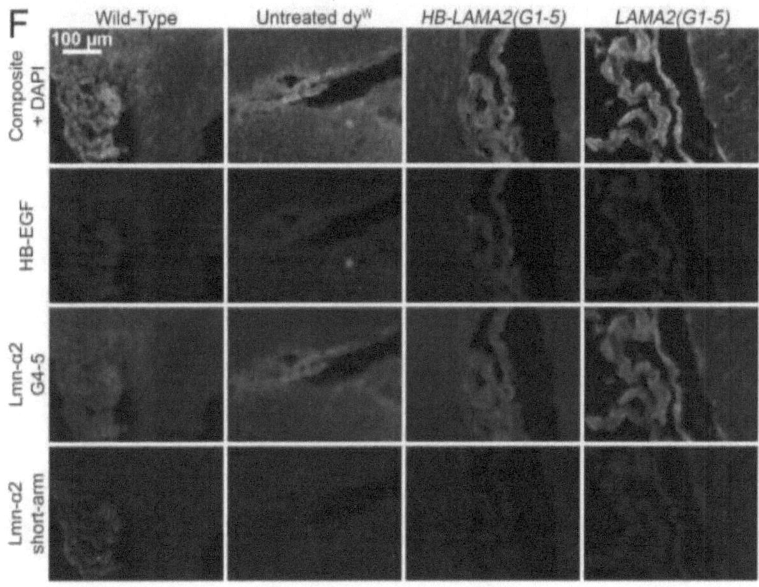

Tissues from dy^W mice treated with intravenous rAAV9 (+), and from untreated (-) dy^W littermates, were collected at 4 months of age. (A) Representative tissue lysates were immunoblotted with antibodies against HB-EGF and human laminin-α2 G5. Anti-GAPDH immunoblotting was performed as a loading control. The 55 kDa band on the laminin-α2 G5 blot represented non-specific secondary antibody binding. Tissue sections from diaphragm (B), heart (C), liver (D), kidney (E), or brain (F) were stained with antibodies against human laminin-α2 G4-5 (Lmn- α2 G4-5, green), HB-EGF (red), laminin-α2 short arm (Lmn- α2 short-arm, far-red or red), and/or integrin-β1 (red). Limited evidence of HB-LAMA2(G1-5) presence in the kidney (F) is shown in the magnified box below the *HB-LAMA2(G1-5)* column.

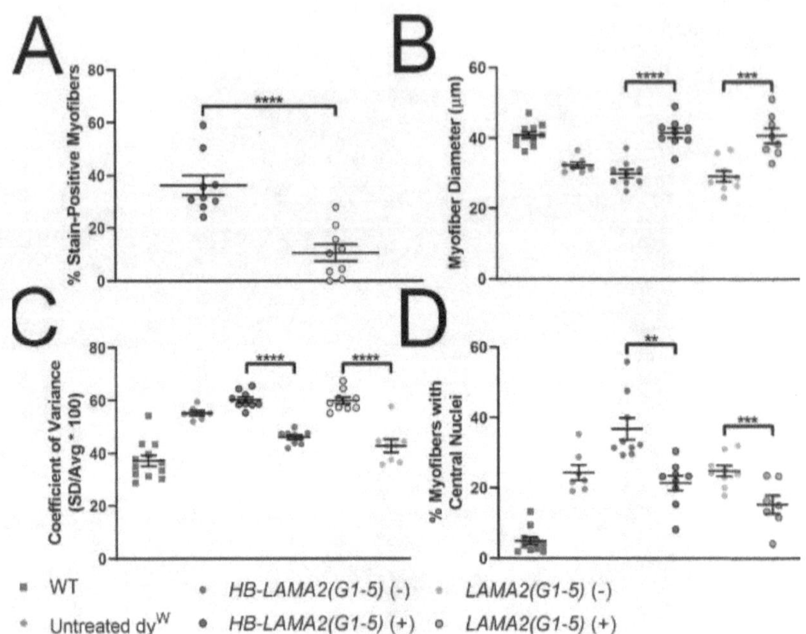

Figure 10 Triceps brachii muscle histopathology after intravenous treatment of dy^W mice with AAV9 expressing micro-laminins

(A) Percentage of myofibers that stained positive for micro-laminin proteins. (B) Minimum (mini-)Feret's diameter of stained myofibers compared to unstained myofibers. (C and D) Quantification of myofiber pathological measures (coefficient of variance in myofiber diameter and percentage of myofibers with central nuclei) in stained and unstained myofibers. Error bars are SEM for n = 9 muscles per condition. **p < 0.01, ***p < 0.001, and ****p < 0.0001.

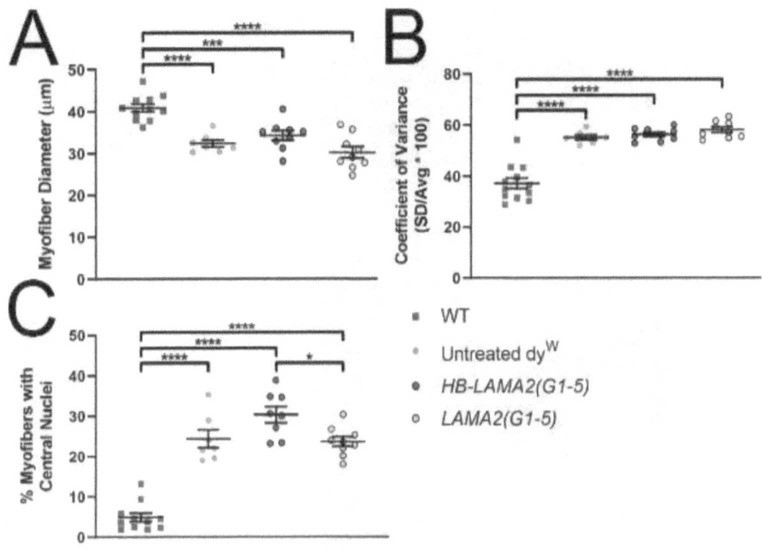

Figure 11 Histopathology dy^W triceps brachii muscles following intravenous rAAV9.micro-laminin injection

Entire triceps brachii muscle sections from 4-month old rAAV9-treated dy^W mice, and from 4-month old untreated wild type and dy^W littermates were analyzed using an automated ImageJ protocol. (A) Myofiber mini-Feret's diameter, (B) Myofiber mini-Feret's diameter coefficient of variance, and (C) Percentage of myofibers with central nuclei were measured. Error bars are SEM for n = 7-12 muscles per condition. *p < 0.05, ***p < 0.001, ****p < 0.0001.

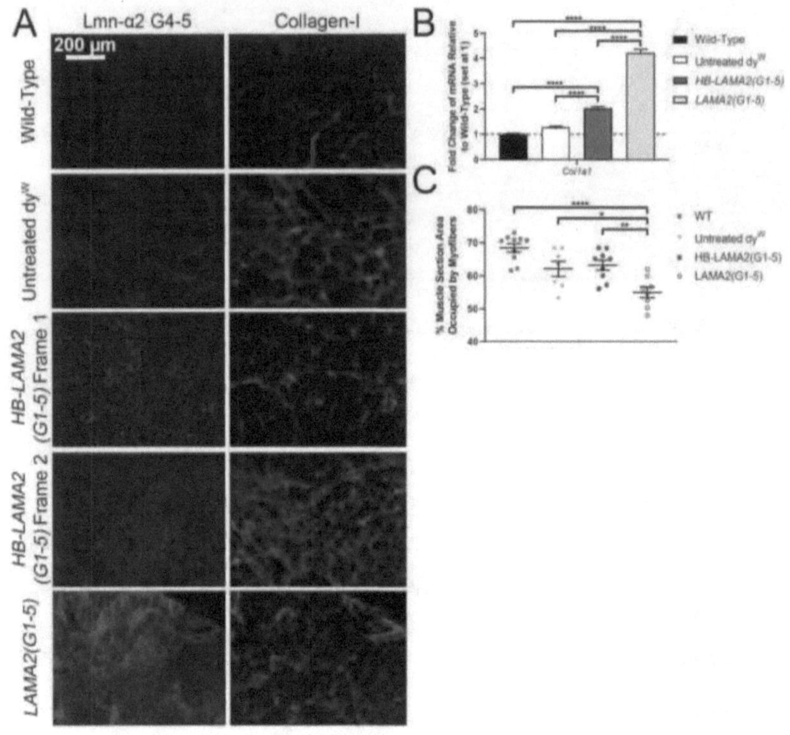

Continued.

Figure 12 Fibrosis and muscle wasting in dy^W muscles following intravenous rAAV9.micro-laminin injection

(A) Muscle sections from mice treated IV with rAAV9 were stained with antibodies against human laminin-α2 G4-5 (green) and collagen-1 (red). Two fields of view were imaged for the *HB-LAMA2(G1-5)* condition to demonstrate the covariation in HB-LAMA2(G1-5) and

Figure 12 continued.

collagen-1 protein expression. (B) Fold-change in *Col1a1* expression was calculated relative to expression in age-matched WT littermates, with 18S ribosomal RNA gene expression as an internal control. (C) The percent of muscle area occupied by myofibers was measured using anti-dystroglycan staining to delineate myofibers. Error bars are SEM for n = 7 - 10 muscles per condition for percent area occupied by myofibers and n = 6 mice per condition for *Col1a1* expression. $*p < 0.05$, $**p < 0.01$, and $****p < 0.0001$.

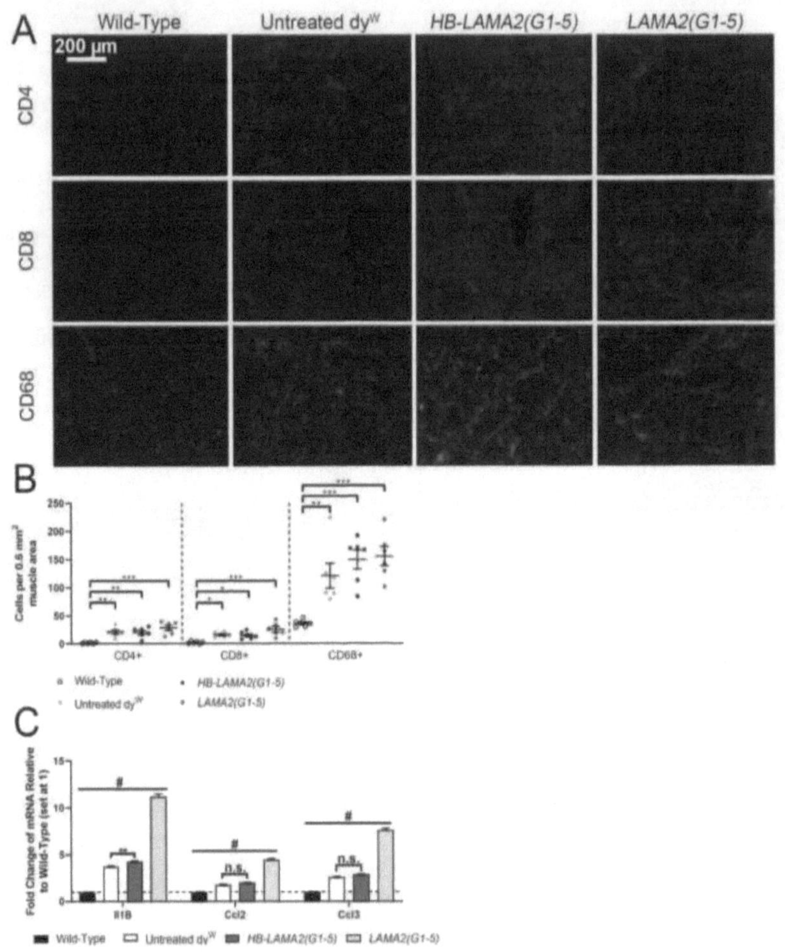

Continued.

Figure 13 Immune cell response in dy^W muscles following intravenous rAAV9.micro-laminin injection

Figure 13 continued.

(A) Muscle sections from mice treated with intravenous rAAV9 were stained with antibodies against CD4, CD8, or CD68. (B) CD4+, CD8+, and CD68+ cells were quantified per 0.6 mm^2 field of view. (C) Fold-changes in gene expression of interleukin β1 (*Ilb1*), C-C chemokine ligand 2 (*Ccl2*, also called MCP1), and C-C chemokine ligand 3 (*Ccl3*, also called MIP1α) were calculated relative to expression in age-matched WT littermates, with 18S ribosomal RNA gene expression as an internal control. Error bars are SEM for n = 6 mice per condition. *p < 0.05, **p < 0.01, ***p < 0.001 and ****p < 0.0001. #p < 0.0001 for all differences unless otherwise stated. n.s. = non-significant.

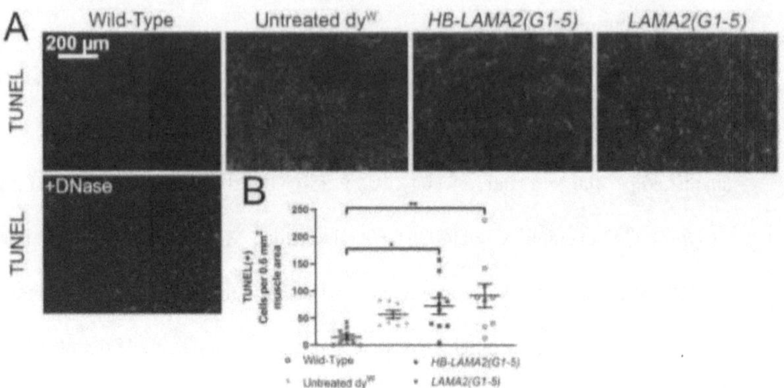

Figure 14 TUNEL staining in dy^W muscles following intravenous rAAV9.micro-laminin injection

(A) Muscle sections from mice treated with intravenous rAAV9 were TUNEL stained (green). A WT muscle section pre-treated with DNase (+DNase) was used as positive control for TUNEL staining. (B) The number of TUNEL(+) nuclei per 0.6 mm^2 field of view was counted using an automated ImageJ macro. Error bars are SEM for 6 mice per condition.

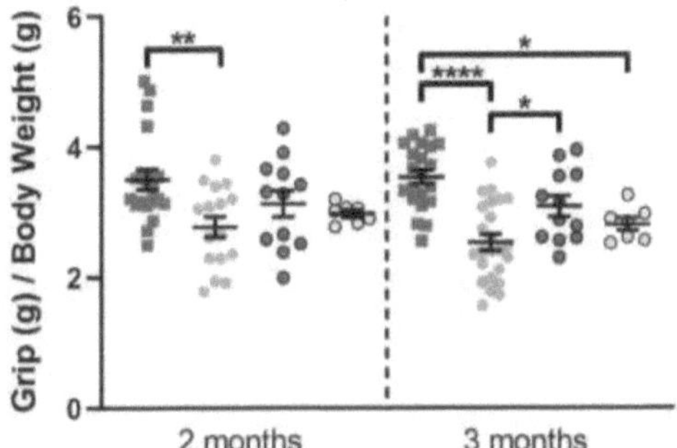

Figure 15 Weight-normalized forelimb grip strength after intravenous treatment of dy^W mice with AAV9 expression micro-laminins

The forelimb grip strength of mice treated with intravenous rAAV9, and that of untreated wild type and dy^W littermates, was assessed at 2 months and at 3 months of age. Grip strength was normalized to body weight for each mouse. Error bars are SEM for n = 7 - 24 mice per condition. *p < 0.05, **p < 0.01, and ****p < 0.0001.

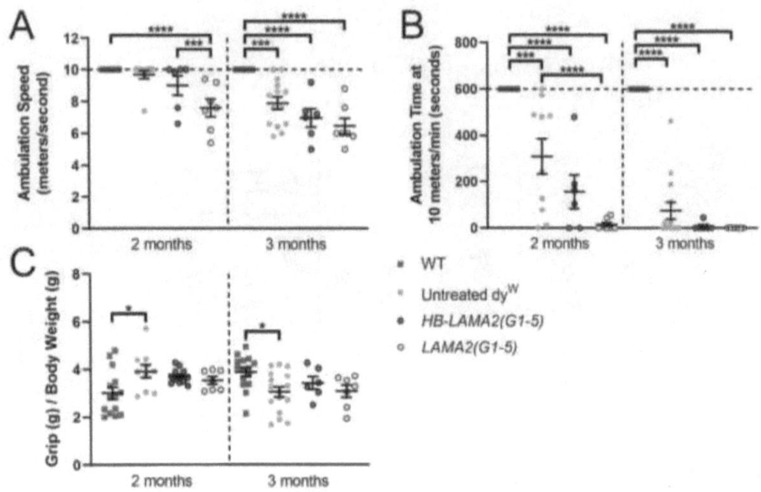

Figure 16 Ambulation and weight-normalized hindlimb grip strength of dy^W mice following intravenous rAAV9.micro-laminin injection

The ambulation speed, ambulation time at 10 meters/minute, and hindlimb grip strength of mice treated with intravenous rAAV9, and that of untreated wild type and dy^W littermates, was assessed at 3 months of age. Grip strength was normalized to body weight for each mouse. Error bars are SEM for n = 6 - 16 mice per condition. *p < 0.05, ***p < 0.001, and ****p < 0.0001.

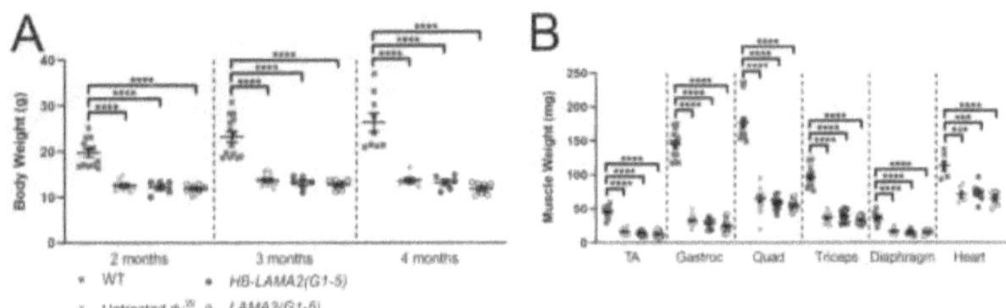

Figure 17 Body weight and muscle weight of dy^W mice following intravenous rAAV9.micro-laminin injection

(A) The body weight of mice treated with intravenous rAAV9, and that of untreated WT and dy^W littermates, was assessed at 2 months, 3 months, and 4 months of age. (B) Wet muscle weight was assessed immediately following dissection at 4 months of age. Error bars are SEM for n = 6 - 14 mice per condition for body weight, n = 12 - 16 muscles per condition for TA, gastrocnemius, quadriceps, and triceps brachii muscles, and n = 6 - 8 per condition for diaphragm and heart. ***p < 0.001 and ****p < 0.0001.

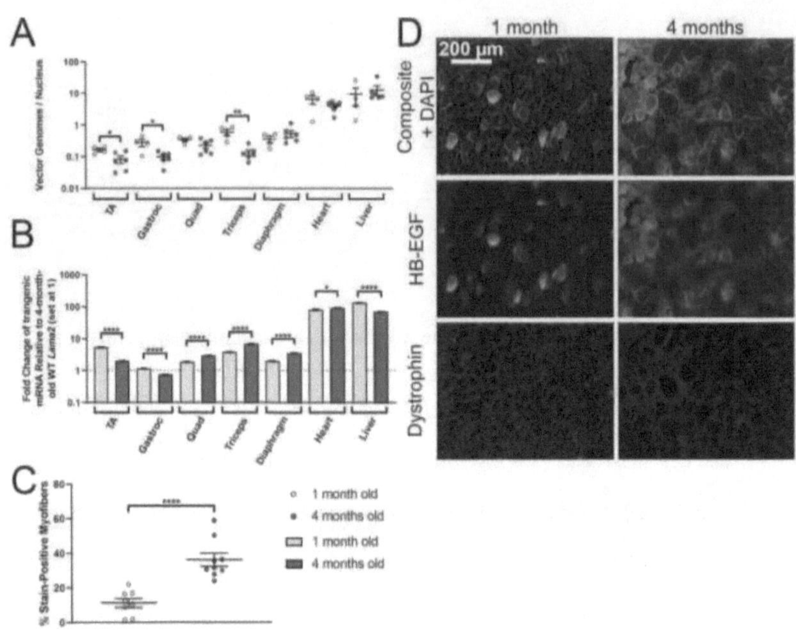

Continued.

Figure 18 Comparison of muscle biodistribution, gene and protein expression at 1 month and at 4 months after IV treatment of dy^W mice with rAAV9.CMV.HB-LAMA2(G1-5) Neonatal (P1) dy^W mice were injected intravenously with 1E+12 vg rAAV9.CMV.*HB-LAMA2(G1-5)*. Tissues were analyzed at 1 month or at 4 months post-injection. (A) The number of AAV genomes per nucleus was quantified in various tissues. (B) Expression of *HB-LAMA2(G1-5)* transgene was determined. Fold change in gene expression was calculated relative to expression of endogenous mouse *Lama2* gene in the corresponding tissues of 4-month old wild type littermates. (C) Percentage of myofibers that stained positively for HB-LAMA2(G1-5) protein. (D) Example of HB-LAMA2(G1-5) staining Figure 18 continued.

using HB-EGF antibody (green) at 1 month and 4 months post-treatment, with dystrophin (red) used as a counterstain to identify myofibers. Composite image includes DAPI (blue) to stain nuclei. Error bars are SEM for n = 4 mice for the 1-month time point and n = 6 mice for the 4-month time point. $*p < 0.05$, $**p < 0.01$, and $****p < 0.0001$.

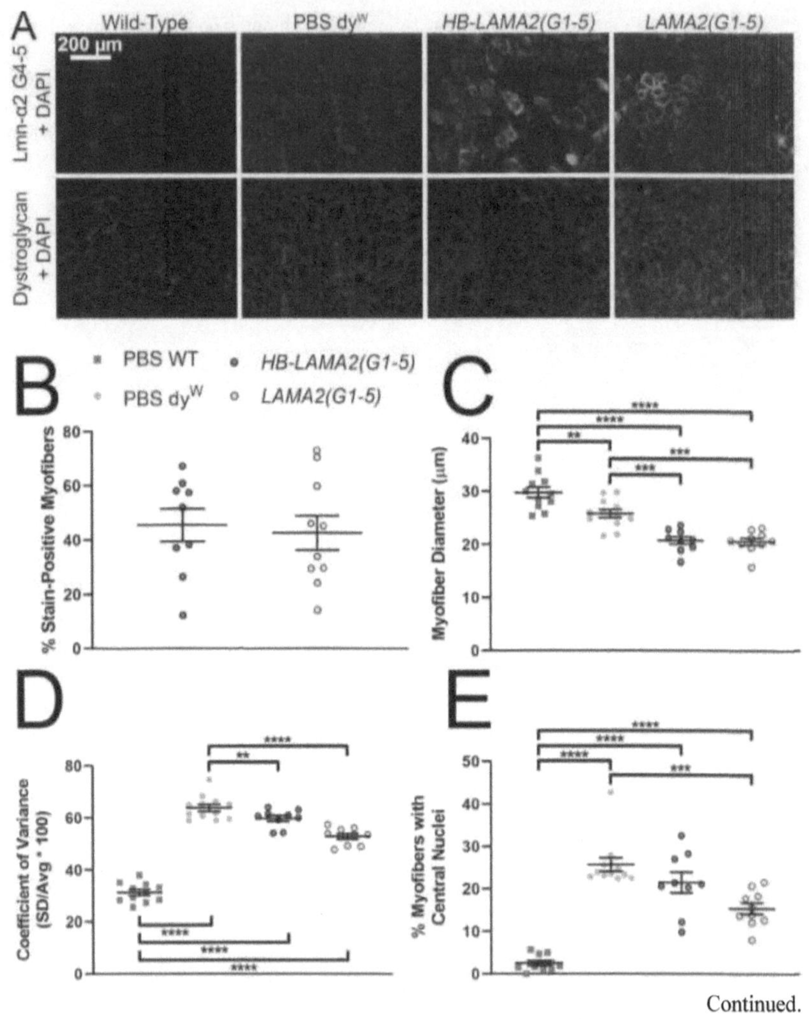

Figure 19 Tibialis Anterior muscle histopathology after intramuscular treatment of dy^W mice with rAAV9 expressing micro-laminin

Figure 19 continued.

The TA muscles of 2-week old dy^W mice were injected with 1E+11 vg rAAV9

containing either *LAMA2(G1-5)* or *HB-LAMA2(G1-5)* genes, or an equal volume of PBS,

and analyzed two months post-injection. (A) Muscle sections stained with antibodies

against human laminin-α2 G4-5 to detect micro-laminins (green), and against

dystroglycan (red) to outline myofibers. (B) Percentage of myofibers that stained

positively for micro-laminins. (B) Myofiber mini-Feret's diameter. (C) Myofiber mini-

Feret's diameter coefficient of variance. (D) Percentage of myofibers with central nuclei.

Error bars are SEM for n = 9 - 12 muscles per condition. **p < 0.01, ***p < 0.001, and

****p < 0.0001.

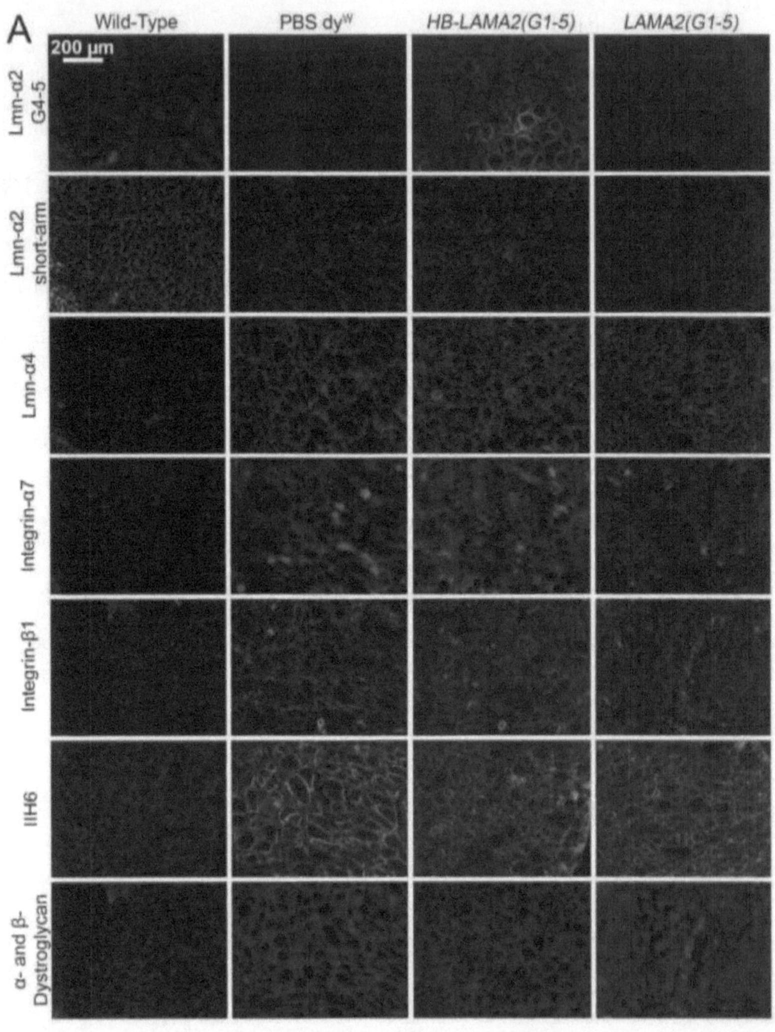

Continued.

Figure 20 Expression of laminins and associated proteins in dy^W muscle following intramuscular rAAV9.micro-laminin injection

Figure 20 continued.

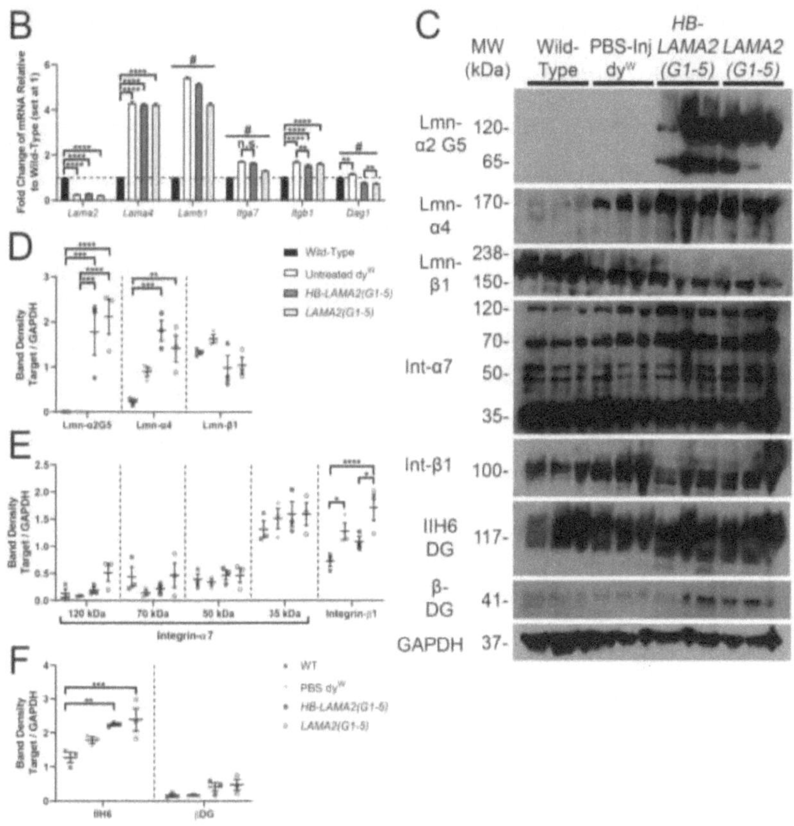

Continued.

Two-week-old dy^W mice were injected with rAAV9 or PBS into the TA muscle and euthanized two months later. (A) Sequential tissue sections were stained with antibodies against human laminin-α2 G5 domain, mouse laminin-α2 short-arm, laminin-α4, laminin-

Figure 20 continued.

α5, integrin-α7, integrin-ß1, dystroglycan laminin-binding glycan (IIH6), and α/ß-dystroglycan. (B) Fold-change in gene expression was calculated relative to expression in age-matched WT littermates, with 18S ribosomal RNA gene expression as an internal control. (C) Muscle lysates were immunoblotted with antibodies, as in (A), with an additional antibody against laminin-ß1. (D-F) Band density was quantified using ImageJ and normalized to GAPDH band density. The 50 kDa doublet in (E) was quantified as a single band. Error bars are SEM for n = 5 - 6 mice per condition for gene expression, and n = 3 mice per condition for immunoblotting. *p < 0.05, **p < 0.01 ***p < 0.001 and ****p < 0.0001. #p < 0.0001 for all differences unless otherwise stated. n.s. = non-significant.

Chapter 3. Engineering matrix-binding trophic factors

3.1 Introduction

Trophic factors generally have short half-lives when secreted into the serum,[206-209] however their diffusion away from treatment-targeted tissues into the circulation, and activity in non-targeted tissues still reduces therapeutic potential, and could have negative consequences.[210-212] The objective of this chapter was to increase the therapeutic potential of insulin-like growth factor 1 (IGF-1) and follistatin (FST) trophic factor gene therapy in muscular dystrophies by improving accumulation and retention in the skeletal muscle ECM.

IGF-1 is normally expressed in the liver,[213, 214] and variant forms are expressed in skeletal muscle following muscle stress or injury.[215-218] IGF-1 binds the receptor tyrosine kinase, insulin-like growth factor receptor-1 (IGFR1),[219-223] with downstream signaling through the MAPK/ERK pathway leading to myoblast proliferation[224-229] and through the PI3K/Akt/mTOR pathway leading to myoblast differentiation and myofiber hypertrophy.[224, 226-228, 230-232] Previous studies found that IGF-1 overexpression induces skeletal muscle hypertrophy,[229, 233-236] reduces atrophy in mouse models of aging,[237, 238] and improves recovery from experimental injury or atrophy,[239-241] . IGF-1 is expressed in dystrophic muscle as well,[242-244] and its overexpression has demonstrated therapeutic efficacy in the mdx model of Duchenne muscular dystrophy,[234, 245-248] and the dystrophic

129 ReJ[249, 250] and dy[W 107, 190, 251] mouse models of laminin-α2-deficient congenital muscular dystrophy.

Follistatin is required for the normal development of the musculoskeletal system.[252-254] It indirectly stimulates muscle growth by binding and sequestering myostatin, a muscle growth inhibitor.[255-260] When not bound by follistatin, myostatin binds the activin receptor IIB (ActRIIB)[261] which, in turn, phosphorylates cytosolic Smad3.[260, 262] p-Smad3 then binds and inhibits Akt of the Akt/mTOR pathway, thereby restricting myofiber hypertrophy,[260] or complexes with Smad4 and translocates to the nucleus where it inhibits transcription of muscle development genes.[263, 264] Follistatin overexpression or myostatin deletion causes myofiber hypertrophy and muscle growth.[257, 258, 261, 265-267] Follistatin may have muscle growth signaling activities beyond inhibiting myostatin, as muscle growth in response to follistatin overexpression can exceed that of myostatin deletion.[257, 258, 266] Recombinant (r)AAV-mediated follistatin overexpression has demonstrated therapeutic efficacy in mouse models of muscular dystrophy,[268, 269] has caused muscle hypertrophy in non-human primates,[270] and has been tested in clinical trials for Becker muscular dystrophy[271] and inclusion body myositis.[272] Follistatin has two isoforms, one with 344 amino acids (aa) precursor to a 315 aa mature protein, and another with 317 aa, precursor to a 288 aa mature protein. The 288 aa protein has some endogenous heparin-binding activity.[273, 274]

To account for short protein half-lives, in order to have long-term effects, trophic factor protein therapy may be administered with repeated injections. While local injections reduce risk of off-target effects, repeated injections are less ideal than a single-treatment therapy. Trophic factor protein therapy may also use an osmotic pump system for continuous administration, however this may increase risk of off-target effects and would still require maintenance. Both protein therapy administration methods would require a large amount of protein to be manufactured, and to remain in continual supply.

rAAV-mediated expression is another way to achieve continual trophic factor administration. Although IGF-1 and follistatin may have salubrious effects in muscle tissue, they may have detrimental effects in other tissues; serum IGF-1 has been associated with prostate cancer,[210] and excess follistatin may disrupt endocrine and reproductive functions.[275, 276] Use of the 344 aa form of follistatin appears to bypass some endocrine signaling issues, including those related fertility, and so the 344 aa form will be used here for these experiments.[271-273, 277] With the goal of increasing trophic factor protein half-life in target tissues, several research groups have fused ECM-binding domains to trophic factors, including IGF-1, finding improved retention and improved trophic factor activity.[278-280] This strategy has not yet been applied to trophic factor gene therapy.

In this chapter, we hypothesized that adding an ECM-binding domain to IGF-1 and to follistatin would increase their accumulation and retention in skeletal muscle ECM, thereby increasing total muscle growth activity and reducing off-target effects. We derived the ECM-binding domain from heparin-binding epidermal growth factor-like growth factor (HB-EGF). The heparin-binding (HB) domain of HB-EGF has strong affinity for heparin sulfate proteoglycans[281] which are a major component of the skeletal muscle ECM.[13, 282, 283] Follistatin has a naturally occurring heparin-binding form

(FST317), derived from alternative splicing,[256, 273] which in initial experiments was tested as well.

3.2 Results

3.2.1 Expression of *HB-IGF1* and *HB-FST* in transfected CHO cells.

The heparin-binding IGF-1 gene, *HB-IGF1*, was designed by adding the human *HBEGF* gene segment encoding the signal peptide (s) through the heparin-binding domain to the 5' end of the human *IGF1* gene segment encoding the mature protein. IGF-1 has several isoforms, distinguished by a C-terminal E-peptide.[213, 284] The isoform most common in muscle, human isoform IGF-1C,[228, 230] was used here. The heparin-binding follistatin gene, *HB-FST*, was designed by adding the same human *HBEGF* gene segment to the 5' end of the human *FST* gene segment encoding the mature 315 aa protein.

Before packaging the recombinant genes in rAAV vectors for animal studies, we asked whether the HB-IGF1 and HB-FST could be properly produced in cultured cells. We chose Chinese Hamster Ovary (CHO) cells as they were sufficient to answer the question and simple to grow and transfect. Upon immunostaining with an anti-IGF-1 antibody (ab9545), CHO cells transfected with pcDNA3.3-TOPO.CMV.*IGF1* displayed cytoplasmic signal, whereas those transfected with pcDNA3.3-TOPO.CMV.*HB-IGF1* displayed both intracellular signal, and a cloud-like extracellular signal surrounding expressing cells, suggesting protein secretion and ECM-binding (Fig. 21A). A similar staining pattern was obtained using a polyclonal anti-HB-EGF antibody (AF-259), for HB-IGF1-expressing cells, while no staining was observed for IGF1-expressing cells. Immunostaining results for pcDNA3.3-TOPO.CMV.*FST* and pcDNA3.3-

TOPO.CMV.*HB-FST* using anti-follistatin (ab47941) and anti-HB-EGF antibodies

mirrored those obtained for *IGF1* and *HB-IGF1* transfections, with HB-FST showing

ECM staining while FST did not (Fig. 21B). Mock-transfected cells did not stain with

any of the antibodies used. These results verified that expression of our transgenes

produced recombinant protein with the correct peptide domains, HB, IGF-1, and

follistatin, and that HB-IGF1 and HB-FST proteins localized differently from their non-

HB counterparts.

**3.2.2 Expression of *HB-IGF1* and *HB-FST* in wild type muscle following rAAV
injection.**

In order to determine the effects of adding a HB-domain to IGF-1 and to FST on

localization and trophic activity in normal skeletal muscle, we injected 1E+11 vector

genomes (vg) bilaterally into the Tibialis Anterior (TA) muscles of adult wild type (WT)

mice and sacrificed the mice for analysis 2 months post-injection. Age-matched control

WT mice were injected with an equal volume of phosphate-buffered saline (PBS). In

these initial experiments, the naturally-occurring heparin-binding form of follistatin,

FST317, was also tested. Vector genome counts were lower than expected given the dose

and route; previous experiments in the laboratory using this method achieved saturation

at greater than 1 vg / nucleus (Fig. 22A). Fold-change in expression was calculated

relative to endogenous mouse *Igf1* or *Fst* expression in the PBS-injected condition (Fig.

22B and C). Expression differed significantly between *IGF1* and *HB-IGF1*, and between

FST and *HB-FST*, possibly due to technical variability in intramuscular (IM) injections.

rAAV9.CMV.*IGF1*-injected muscles produced an intracellular staining pattern in myofibers with an anti-IGF-1 antibody, and no signal with an anti-HB-EGF antibody (Fig. 23A). rAAV9.CMV.*HB-IGF1*-injected muscles, on the other hand, produced an ECM staining pattern with an anti-IGF-1 antibody, co-localizing with an anti-HB-EGF antibody signal. rAAV9.CMV.*FST*-injected muscles displayed punctate staining with an anti-follistatin antibody, largely within interfascicular spaces (Fig. 23B). Both rAAV9.CMV.*FST317*- and .*HB-FST*-injected muscles produced ECM staining with the anti-follistatin antibody, staining which colocalized with anti-HB-EGF antibody staining in the *HB-FST* condition. Thus, the presence of the HB domain appeared to confer improved ECM localization upon both trophic factors.

3.2.3 HB-IGF1 and HB-FST protein accumulation in WT skeletal muscle following IM rAAV injection.

We next performed western blots to characterize the recombinant proteins (Fig. 23C). Anti-HB-EGF blotting revealed bands at 15 kDa for the HB-IGF1-expressing muscle protein lysates and 50 kDa for the HB-FST-expressing muscle lysates, close to expected molecular weights for both full-length proteins. Both bands were remained following heparin-agarose (H-A) precipitation. Anti-follistatin blotting revealed a 50 kDa band for the HB-FST condition, which remained following H-A precipitation, and a 38 kDa band for the FST condition which was absent following H-A precipitation. Twice the microgram amount of protein lysate had to be loaded from FST-expressing muscle than from HB-FST-expressing muscle to detect a band, suggesting increased concentration of

HB-FST protein compared to recombinant follistatin protein in their respective lysates, although *HB-FST* also had greater mRNA expression than *FST* (Fig. 23C).

Anti-IGF-1 blotting revealed two pairs of bands in the *HB-IGF1* condition, one at 37 kDa and 32 kDa, and the other at 20 kDa and 15 kDa. Each of these bands remained following H-A precipitation. Anti-IGF-1 blotting in the *IGF1* condition found a single band at 30 kDa and a pair of bands at 16 kDa and 14 kDa. H-A precipitation retained each band from the *HB-IGF1* condition, but only the 14 kDa band from the *IGF1* condition. Again, twice the microgram amount of protein lysate had to be loaded from the *IGF1* condition than from the *HB-IGF1* condition to detect a band, suggesting that HB-IGF1 protein accumulated more in treated muscle than IGF-1 protein.

One hypothetical explanation for the band multiplicity observed in the *HB-IGF1* and *IGF1* conditions is that the 14 kDa and 16 kDa band for the IGF1 condition represent post-translational processing of IGF-1, while the 30 kDa band represents IGF-1 bound to a carrier protein, such as IGF-1 binding protein (IGF1BP). The 32 kDa band, present in the HB-IGF1 condition, would correspond to processed HB-IGF1 bound to a carrier protein, by either the endogenous heparin-binding domain, or the novel HB-domain, but still able to bind heparin-agarose during precipitation via its non-IGF1BP-bound domain. Another possible explanation for the higher molecular weight species would be oligomerization of HB-IGF1 that was resistant to denaturation buffer.

3.2.4 Effects of rAAV-mediated *HB-FST* or *HB-IGF1* expression in WT mouse muscle.

Muscle weight and myofiber diameter data were sex-stratified for WT mice, as these metrics normally vary significantly between male and female mice. Intramuscular injection of rAAV.CMV.*FST* in the TA muscles yielded significant increases in muscle mass and myofiber diameter (122.3 ± 4.8 mg and 54.9 ± 2.9 μm, n = 6 muscles from 3 male mice, and 90.1 ± 6.4 mg and 49.9 ± 1.6 μm, n = 6 muscles from 3 female mice), relative to PBS-injected control muscles (65.7 ± 3.8 mg, p < 0.0001 and 41.7 ± 1.7 μm, p = 0.0014, n = 6 muscles from 3 male mice, and 39.9 ± 2.1 mg, p < 0.0001 and 34.6 ± 0.9 μm, p < 0.0001, n = 6 muscles from 3 female mice) (Fig. 24A and B). Notably, this increase was not seen in the rAAV.CMV.*HB-FST*-injected muscle. Neither rAAV.CMV.*IGF1*-injected nor rAAV.CMV.*HB-IGF1*-injected TA muscle mass was significantly greater than PBS-injected control muscle.

In order to assess changes in signal transduction pathways relevant to IGF-1 and follistatin signaling, we compared protein and phospho-protein western blots for Akt, mTOR, ERK42 and ERK44 (Fig. 25). The MAPK/ERK pathway is relevant primarily to IGF-1 signaling, [224-229] while the PI3K/Akt/mTOR pathway is involved in both IGF-1 and follistatin signaling. [224, 226-228, 230-232, 260] By densitometry, there were no significant changes in PI3K/Akt/mTOR signaling, assayed by phospho-Akt and phospho-mTOR, nor in MAPK/ERK pathway, assayed by phospho-ERK44/42, between any conditions. Thus, expression of IGF1, HB-IGF1, FST, and HB-FST did not appear to stimulate a significant increase in signaling through these kinases. FST signaling via inhibition of myostatin activation of the Smad2/3 pathway was not evaluable (data not shown).

3.2.5 HB-IGF1 and HB-FST expression in dy^{W} mice following IM rAAV injection

IGF-1 is better known for promoting muscle growth in damaged muscle than in healthy, adult muscle. In order to better detect possible differences in function between HB- and non-HB-trophic factor expression, we therefore repeated the IM injection experiments in young dystrophic, laminin-α2-deficient dy^W mice, including PBS-injected WT and dy^W littermate control conditions. The dy^W mouse model was chosen for its early onset and severe symptoms. In this case, vg levels in injected muscle averaged above saturation in each rAAV condition, with at least 1 vector genome being present per mouse muscle cell nucleus (Fig. 26A). As seen in WT IM injections, *IGF1* gene expression was greater than *HB-IGF1* gene expression (Fig. 26B), while *HB-FST* gene expression was greater than *FST* gene expression (Fig. 26C). Fold-changes were calculated relative to mouse *Igf1* or mouse *Fst* in PBS-injected WT littermate muscle.

Anti-IGF-1 antibody staining identified expression of IGF-1 within myofibers and identified HB-IGF1 staining coincident with the ECM outside of skeletal myofibers; anti-HB-EGF staining overlapped with anti-IGF-1 staining in the HB-IGF1-treated condition (Fig. 27A). Anti-follistatin staining was confined to small regions of the myofiber membrane in both FST- and HB-FST-expressing myofibers, though staining was more abundant in the HB-FST condition. Again, anti-HB-EGF staining overlapped with follistatin staining in HB-FST-treated muscles (Fig. 27B). In all dy^W muscles, there was residual laminin-α2 present by short-arm staining, as expected,[73] though expression was less abundant and more variable than found in WT muscle.

3.2.6 Effects of rAAV-mediated *HB-IGF1* expression on dy^W mouse muscle following IM injection.

TA muscle weights from PBS-injected WT mice were significantly greater than TA muscle weights from all dy^W conditions (Fig. 28A). rAAV.CMV.*HB-IGF1*-injected dy^W TA muscle weight (26.9 ± 2.7 mg, n=12 muscles, 6 mice) was significantly greater than PBS-injected dy^W muscle weight (17.3 ± 0.9 mg, n = 10 muscles, 5 mice, p = 0.032). As with muscle weight, average myofiber diameter was greater in WT mice than in dy^W mice of all conditions (Fig. 28B), while markers of pathology, myofiber diameter variation (Fig. 28C) and central nucleation (Fig. 28D), were lower. Notably, average myofiber diameter for the HB-IGF1 condition was not greater than that for the PBS dy^W condition, at odds with the increase seen in HB-IGF1 muscle weight.

Since myofiber hypertrophy (increase in myofiber diameter), did not explain the mass gain in the *HB-IGF1* condition over the PBS dy^W condition, we asked whether an increase in myofiber number, or a reduction in myofiber death was responsible. First, activation of PI3K/Akt/mTOR and MAPK pathways was assayed by phosphoprotein immunoblot (Fig. 29A). Akt phosphorylation, most pertinent to myofiber hypertrophy,[240, 285] was not increased in rAAV treatment conditions over PBS control conditions, although total levels of Akt protein appeared increased in dy^W muscle relative to WT levels (Fig. 29B). Activation of the MAPK pathway, relevant to cell proliferation,[228] also was not increased in treatment conditions relative to the control PBS-injected condition, though again total levels of protein appeared increased in dy^W muscle over WT muscle (Fig. 29C). All dy^W muscle had increased cell proliferation by Ki67 staining relative to WT muscle, however there were no differences between rAAV-treated and PBS-treated dy^W muscle (Fig. 30A). High background impeded cell death quantification by TUNEL

staining; neither rAAV treatment reduced TUNEL-positive nuclei per field of view (fov;
Fig. 30B). In fact, TUNEL-positive nuclei were greater in the *IGF1* condition (39.6 ± 5.8
nuclei / fov, n = 4 muscles) than in the *HB-IGF1* condition (14.9 ± 2.6 nuclei / fov, n = 4
muscles, p = 0.0024) and PBS-injected control dy^W mice (23.2 ± 1.2 nuclei / fov, n = 4
muscles, p = 0.0320).

3.2.7 HB-IGF1 and HB-IGF1-LAMA2(G1-5) expression in dy^W muscle following IV rAAV injection.

Since neither rAAV9.CMV.*FST*, nor rAAV9.CMV.*HB-FST* demonstrated benefit
in dy^W mouse muscle after IM injection, testing of these treatments was at this point
abandoned. An additional treatment, rAAV.CMV.*HB-IGF1-LAMA2(G1-5)* was added, in
which the HB-IGF1 protein is supplied with a C-terminal laminin-a2 G1-5 domain. This
protein is intended to provide both the trophic support of IGF-1, and the structural
support of HB-LAMA2(G1-5) assayed in Chapter 2. HB-IGF1-LAMA2(G1-5) protein
expression was first assayed in WT mice following IM rAAV injection and compared to
expression of HB-EGF, IGF-1, HB-IGF1, LAMA2(G1-5), and HB-LAMA2(G1-5)
proteins. HB-IGF1-LAMA2(G1-5) was found to have the expected sarcolemmal staining
pattern and immunoreactivity with antibodies against HB-EGF, IGF-1, and human
laminin-α2 G4-5 (Fig. 31A).

On anti-HB-EGF immunoblot, rAAV9.CMV.*HB-IGF1-LAMA2(G1-5)*-injected
muscle lysate displayed a band around 150 kDa, near the HB-LAMA2(G-15) band as
well as several fainter bands, likely cleavage fragments (Fig. 31B). Both the 150 kDa and
cleavage fragment bands were retained following heparin-agarose precipitation. Both

anti-IGF-1 and anti-laminin-a2 G5 immunoblotting also produced the 150 kDa band in the HB-IGF1-LAMA2(G1-5) muscle lysate lane.

For this IV experiment, data from IV rAAV.CMV.*HB-LAMA2(G1-5)* tests was included for comparison. As in Chapter 2, IV experiments were complicated by low transduction rates in skeletal muscle, well below the saturation point of 1 vg per nucleus (Fig. 32A). Transgenic mRNA expression levels varied greatly between conditions, with consistently lower expression for *IGF1*, and higher expression for *HB-IGF1-LAMA2(G1-5)* except in the diaphragm, liver, and kidneys (Fig. 32B).

In general, few myofibers stained positively for recombinant proteins. However, differences in staining patterns were evident between cells expressing each recombinant protein. In both the triceps brachii muscles (Fig. 33A) and the heart (Fig. 33B), IGF-1 protein displayed primarily intracellular staining, whereas both HB-IGF1 and HB-IGF1-LAMA2(G1-5) displayed sarcolemmal staining. As expected, muscle tissue from the *IGF1* condition displayed immunoreactivity only toward the anti-IGF-1 antibody, while tissue from the *HB-IGF1* condition displayed additional immunoreactivity with an anti-HB-EGF antibody, and tissue from *HB-IGF1-LAMA2(G1-5)* displayed additional immunoreactivity with an anti-human laminin-α2 G4-5 antibody.

3.2.8 Effects of neonatal IV rAAV treatments on dy^W muscle.

Average myofiber diameters were lower, and markers of muscle pathology, myofiber diameter variation and myofiber central nucleation, were greater in both treated and untreated dy^W mouse triceps brachii muscle relative to WT muscle (Fig. 34A-C). Muscle wasting, quantified by the percentage of muscle area occupied by myofibers, was greater in treated and untreated dy^W mice than in WT littermates (Fig. 34D). None of the tested rAAV treatments improved these histological measurements, however this result was confounded by the low percentage of myofibers expressing recombinant protein (Fig. 34E). Within triceps brachii muscle of treated mice alone, myofiber diameter was greater for HB-IGF1-LAMA2(G1-5)-positive myofibers (40.77 ± 2.02 µm, n = 5 muscles) than for negative myofibers (27.22 ± 0.54 µm, n = 5 muscles, p = 0.0002 ; Fig. 34F). Diameter variation was lower in both HB-IGF1-positive myofibers (36.17 ± 3.64 µm/µm, n = 5) and HB-LAMA2(G1-5)-positive myofibers (44.63 ± 1.45, n = 5) than for negative myofibers within the same muscle (58.87 ± 1.62 µm/µm, n = 5 muscles, p = 0.0002 and 57.46 ± 1.90 µm/µm, n = 6 muscles, p = 0.0004, respectively; Fig. 34G).

3.2.9 Effects of neonatal IV rAAV treatments on dy^W behavior.

For behavioral measurements, WT and dy^W data points were consolidated between previous tests in chapter 2 and new tests for this chapter. Body weight-normalized forelimb grip strength was significantly reduced in dy^W mice relative to WT littermates at 2 and 3 months of age (Fig. 35A). At 3 months of age, this measurement was significantly greater in rAAV.CMV.*IGF1*-treated mice (3.65 ± 0.19 g/g, n = 3 mice) than in untreated dy^W mice (2.78 ± 0.14 g/g, n = 18 mice, p = 0.0348), despite low transduction levels in rAAV.CMV.*IGF1*-treated mice. No other treatment significantly improved forelimb grip strength. Due to hindlimb contractures, which cause the hindlimbs of dy^W mice to stiffen against the grid of the grip strength apparatus, the weight-normalized hindlimb "grip" strength of dy^W mice was not significantly reduced relative to WT mice (Fig. 35B). None of the treatments improved hindlimb grip strength relative to untreated dy^W mice. Maximum ambulation speed achieved at two months of age was not significantly different between treated and untreated dy^W mice, and WT littermates (Fig. 35C). At 3 months of age, dy^W mice had lower average maximum ambulation speed relative to WT mice, but none of the treatment conditions improved this measurement. Ambulation time at 10 meters / min was significantly decreased in dy^W mice relative to WT mice at both 2 months and 3 months of age, and none of the treatment conditions improved this measurement (Fig. 35D).

3.2.10 Effects of neonatal IV rAAV treatments on dy^W body and muscle weights.

Both untreated and treated dy^W mice were smaller than their WT littermates at 2, 3, and 4 months of age, and treatment did not improve body weight (Fig. 36A). Likewise, skeletal muscles and heart were smaller in untreated and treated dy^W mice relative to WT littermates (Fig. 36B). Unlike IM injection, HB-IGF1 expression did not improve skeletal muscle weight, nor did any other treatment condition.

3.3. Discussion

While trophic factor therapy has proven beneficial in animal models of muscular dystrophy,[107, 190, 234, 245-251, 268, 269] retention in tissue is problematic due to protein half-life[206, 207] and potential secretion into the circulation. In this Chapter, we attempted to enhance trophic factor gene therapy by adding an ECM-binding domain to muscle trophic factors IGF-1 and follistatin. We hypothesized that the addition of an ECM-binding domain would enhance the muscle growth effects and reduce off-target effects of these two factors by increasing their accumulation and retention within the skeletal muscle

ECM. This strategy has been used previously in protein therapy,[278-280] but not in gene therapy.

In culture transfection experiments, HB-IGF1 and HB-FST protein demonstrated novel ECM-localization (Fig. 21). In both IM and IV injection routes, expression of *HB-IGF1* and *HB-FST* demonstrated new sarcolemmal localization within skeletal muscle relative to their native-form counterparts (Figs. 23, 27, 31, 33). As hypothesized, the addition of an ECM-binding domain increased accumulation of HB-IGF1 and HB-FST protein in skeletal muscle relative to IGF-1 and follistatin, respectively (Fig. 23).

As expected from previous studies,[257, 258, 261, 265-267] FST overexpression resulted in increased muscle weights and myofiber diameters in rAAV-injected WT muscle. Of note, HB-FST expression failed to increase muscle weight. Future studies will test whether this was due to possible structural changes induced by the additional HB domain that may inhibit the myostatin-binding site of follistatin.

While follistatin was superior to other conditions in increasing WT mouse muscle weights, it failed to increase muscle weight in dy^W mice following IM rAAV injection. HB-IGF1 expression, instead was the only condition to increase dy^W muscle weight following IM rAAV injection. This increase was apparently not due to myofiber hypertrophy, hyperplasia, or protection from apoptosis (Figs. 29 and 30), though the apoptosis results were complicated by high background TUNEL staining. One hypothesis why HB-IGF1, and IGF-1 in previous studies in dy^W mice,[107, 190, 251] but not follistatin, were able to induce muscle growth in dy^W mice involves differences in IGF-1 and follistatin signaling. Dy^W muscles exhibit reduced regeneration,[76, 78] which may be due to reduced cell adhesion signaling involving Akt activity.[33] While follistatin indirectly

increases Akt activity by inhibiting its inhibition,[266] IGF-1 directly increases Akt

activity[224, 226-228, 230-232] and might therefore overcome a lack of cell adhesion signaling.

Despite improvements in muscle weight seen in IM rAAV.CMV.*HB-IGF1*-

injected dy^W muscle and previous evidence of IGF-1 expression improving outcome in

dy^W mice,[107, 190, 251] neither *HB-IGF1*, *IGF1*, nor *HB-IGF1-LAMA2(G1-5)* vectors

improved muscle weight following IV rAAV injection (Fig. 36). Within treated muscle,

however, HB-IGF1-LAMA2(G1-5)-expressing myofibers were larger and had less

variation in diameter than non-expressing myofibers (Fig. 34F). With the exception of

rAAV.CMV.*IGF1* increasing weight-normalized forelimb grip strength, behavioral

measurements were unaffected by rAAV treatments (Fig. 35).

Due to the low transduction rates in IV rAAV-injected mice, we were unable to

evaluate secretion of recombinant protein into the circulation. This second arm of the

hypothesis is important to determine, as off-target effects following escape of trophic

factors into the circulation is a major concern for trophic factor gene therapy. Additional

future experiments may determine whether the lack of muscle growth activity in HB-FST

is due to loss of myostatin-binding activity, or to some other difference between HB-FST

and follistatin proteins, and the reason for follistatin's lack of growth activity in dy^W

mice.

These studies demonstrate that attaching an ECM-binding domain to trophic

factors may change their localization, and accumulation within targeted tissues. Activity

may either be retained, as with HB-IGF1 or lost, as with HB-FST. Although it was not observed in these gene therapy studies, the addition of ECM-binding domains improved trophic factor activity in protein therapy,[278-280] and it stands to reason that, with additional modifications, an improvement may be seen in trophic factor gene therapy.

3.4 Materials and Methods

3.4.1 Production of AAV9 vectors containing trophic factor genes

HB-IGF1 was designed using the human *HBEGF* gene segment encoding the signal peptide through heparin-binding domain from NM_001945.2 (base pairs 276 to 597), and the human *IGF1* gene segment encoding the mature peptide from NM_001111283 (base pairs 328 to 741). *HB-IGF1-LAMA2(G1-5)* was designed as *HB-IGF1* with the addition of the human *LAMA2* gene segment encoding the G domains from RefSeq NM_000426.3 (base pairs 6538 to 9436). *HB-FST* was designed using the same *HBEGF* gene segment and the human *FST* gene segment encoding the mature peptide from NM_013409.2 (base pairs 446 to 1390). The recombinant genes were synthesized by GeneArt Gene Synthesis (Regensburg, Germany) in a pMA-T vector backbone. The genes were subcloned into a pcDNA3.3-TOPO backbone for use in cell culture transfections, and into a rAAV plasmid with CMV promoter, SV40 enhancer, and SV40 poly-adenylation signal for rAAV vector production. From 5' ITR to 3' ITR, the genome length of rAAV.CMV.*IGF1* was 1.881 kb, the length of rAAV.CMV.*HB-IGF1* was 2.133 kb, the length of rAAV.CMV.*HB-IGF1-LAMA2(G1-5)* was 4.836 kb, the length of rAAV.CMV.*FST* was 2.439 kb, the length of rAAV.CMV.*HB-FST* was 2.673 kb, and the length of rAAV.CMV.*FST317* (from NM_006350.3) was 2.352 kb. Plasmids were packaged in rAAV9 vectors by the Nationwide Children's Hospital Viral Vector Core

(Columbus, OH) using the triple transfection technique in HEK293 cells,[130] and highly purified using sucrose density centrifugation and anion exchange chromatography, as previously described.[202]

3.4.2 Cell Transfection

CHO-K1 cells (ATCC, Manassas, Virginia) were grown in 10 cm dishes in DMEM (Thermo Fisher Scientific) with 10% fetal bovine serum (Millipore-Sigma). When at 80% confluence, cells were transfected with pcDNA3.3-TOPO.CMV.*IGF1*, .*HB-IGF1*, .*FST*, or .*HB-FST* using TransIT-X2 Dynamic Delivery System (Mirus, Madison, Wisconsin), according to the manufacturer's instructions in 10 cm dishes.

3.4.3 Mice

All mouse experiments were performed using protocols approved by the Institutional Animal Care and Use Committee at The Abigail Wexner Research Institute at Nationwide Children's Hospital (Columbus, OH). C57BL/6J mice were purchased from Jackson Laboratories (Bar Harbor, ME). Dy^W mice bred on a pure C57BL/6 background were a generous gift from Dr. Dean Burkin at University of Nevada, Reno, Medical School.

3.4.4 Intramuscular AAV Injections

Adult mice were anesthetized with isofluorane (Baxter, Deerfield, IL), and hindlimbs were shaved over the TA muscles. 1E+11 vg rAAV9 in 25 μL PBS was injected into each TA muscle using a 0.33cc Thinpro Insulin Syringe (Terumo Medical, Somerset, NJ). Two-week-old mice were restrained and 1E+11 vg rAAV9 in 5 μL PBS was injected into each TA muscle using a 0.3cc Thinpro Insulin Syringe.

3.4.5 Intravenous AAV Injections

Mouse pups up to 48 hours old were anesthetized on ice and injected with 1E+12 vg rAAV9 into the temporal vein using a Thinpro 0.3cc Insulin Syringe. The minimum injection volume allowable by virus titer was used.

3.4.6 Grip Strength

On three consecutive days, one week prior to grip strength assessment, mice were trained on the Grip Strength Meter (Columbus Instruments, Columbus, OH). Each training day consisted of five forelimb and five hindlimb grip strength trials which were not recorded. The week of assessment, grip strength was assessed on five consecutive days. Each assessment consisted of five forelimb trials using the "T Peak" setting and five hindlimb trials using the "C Peak" setting on the Grip Strength Meter and measured in grams, followed by body weight measurement. For each mouse, on each day of the assessment week, the five trials were averaged and normalized to body weight obtained on that day. For each mouse, the final grip strength value was the average of the body weight-normalized grip strength values obtained over the five days.

3.4.7 Ambulation

On three consecutive days, one week prior to ambulation assessment, mice were trained on the Treadmill Meter (Columbus Instruments, Columbus, OH). The week of assessment, mice were placed on the stationary Treadmill Meter, one or two mice per enclosed lane. The treadmill was started at 5 meters per minute, increasing 1 meter per minute every 30 seconds until 10 meters per minute was reached. The treadmill was run at 10 meters per minute for 10 minutes. For each mouse, the trial ended when the mouse fell off the end of the treadmill three times in a span of 20 seconds. At the end of the trial, the maximum speed reached and time of ambulation at 10 meters per minute were recorded. If a mouse failed to reach 10 meters per minute, 0 seconds was recorded for ambulation time. For each mouse, the final values for speed and ambulation time were the averages of those obtained over the five days.

3.4.8 Tissue Processing

Skeletal muscles were removed, weighed, mounted with O.C.T. (Thermo Fisher Scientific) on cork, and snap-frozen liquid nitrogen-cooled 2-methylbutane. The internal organs and heart were removed, placed in Peel-A-Way Disposable Embedding Molds (Thermo Fisher Scientific), covered in Tissue Plus O.C.T. (Thermo Fisher Scientific) and frozen in a 2-methylbutane (Thermo Fisher Scientific) / dry ice bath.

3.4.9 AAV Biodistribution

DNA was extracted from tissue shavings using a DNeasy Blood & Tissue Kit (Qiagen, Germantown, MD). Vector genome quantity in 100 ng of sample DNA was determined using 2X TaqMan Gene Expression Master Mix, a standard curve of linearized rAAV plasmid from 5 vg to 5E+6 vg, and the following primer/probe set amplifying a region within the CMV promoter: 5'-CTAACGCCAATAGGGACTTTCC-3' (forward), 5'- 56-FAM/ACGGTAAAC/ZEN/TGCCCACTTGGCAGTACA/3IABkFQ/-3' (probe), and 5'-GGGCGTACTTGGCATATGATAC-3' (reverse). Vector genomes per mouse nucleus were then calculated using an estimate of 1.7E+5 mouse nuclei per 1 µg of genomic DNA.

3.4.10 Gene Expression Measurement

RNA was extracted from tissue shavings using a Direct-zol RNA MiniPrep kit (Zymo

Research, Irvine, CA), and cDNA was prepared using a High-Capacity cDNA Reverse

Transcriptase kit (Applied Biosystems, Bedford, MA). Transgene expression was

measured in cDNA samples using TaqMan Gene Expression Master Mix and *IGF1*

(human) Assay Hs01547656_m1 (Applied Biosystems) for *IGF1* and *HB-IGF1*

transgenes, or *FST* (human) Assay Hs01121164_m1 (Applied Biosystems) for *FST* and

HB-FST transgenes. Additional assays included mouse *Igf1* Assay Mm00439560_m1 and

mouse *Fst* Assay Mm00514982_m1 (Applied Biosystems). 18S rRNA was used as an

internal control (Applied Biosystems). The $2^{-\Delta\Delta Ct}$ method was used to calculate fold-

change in expression.[286]

3.4.11 Heparin-Agarose Precipitation

200 µL Heparin-Agarose Type I, saline suspension (Millipore-Sigma) was added per mL

of cell lysate or muscle lysate to be precipitated, and mixed overnight at 4°C. Precipitated

samples were washed thoroughly with PBS. Precipitated samples were resuspended in 50

µL 1X NuPage LDS Sample Buffer with 100 mM β-mercaptoethanol. 20 µL of the

resuspension was boiled and used for immunoblotting.

3.4.12 Immunoblotting

Protein was extracted from tissue shavings using NP40 lysis buffer: 1% (w/v) NP40

(Millipore-Sigma, St. Louis, MO) in 50 mM Tris buffer with 150 mM NaCl, 1 mM

EDTA (Invitrogen, Grand Island, NY), and cOmplete EDTA-free Protease Inhibitor

Cocktail (Millipore-Sigma). 40 µg of each protein lysate was added to NuPAGE LDS

Sample Buffer (Thermo Fisher Scientific) with 100 mM β-mercaptoethanol (Thermo

Fisher Scientific) and boiled for 10 minutes prior to gel electrophoresis and transfer to

nitrocellulose membranes for immunoblotting. Antibodies included anti-HB-EGF (AF-

259; R&D Systems, Minneapolis, MN), anti-IGF-1 (ab9572) and anti-follistatin

(ab47941; Abcam, Cambridge, UK), anti-laminin-α2 G5 (2D4; Abnova, Taipei, Taiwan),

and anti-GAPDH (G9545; Millipore-Sigma, Burlington, MA). The HRP-conjugated

secondary antibodies against goat and rabbit IgG were obtained from Jackson

ImmunoResearch (West Grove, PA), and the secondary antibody against mouse IgG1

was obtained from Bethyl Laboratories (Montgomery, TX). Protein band densities were

measured using ImageJ 1.46r.

3.4.13 Immunofluorescent Staining

10 μm tissue sections, cut on a Cryostat, were placed on Fisher Superfrost Plus glass

slides (Thermo Fisher Scientific). A hydrophobic border was drawn around each section

using an ImmEdge Hydrophobic Barrier PAP pen (Vector Labs, Burlingame, CA).

Sections were fixed in 2% Paraformaldehyde (EMS) in PBS for 10 minutes,

permeabilized in 0.1% Triton X-100 (Millipore-Sigma) in PBS for 10 minutes, incubated

in Mouse-on-Mouse Blocking Reagent (Vector Labs) for 1 hour, and blocked with 10%

donkey serum in PBS for 30 minutes prior to incubation overnight at 4°C in primary

antibodies in blocking buffer. Sections were then incubated with secondary antibodies in

blocking buffer for 1 hour at room temperature and then treated with ProLong Gold with

DAPI (Invitrogen) before adding a coverslip. Antibodies included anti-HB-EGF (AF-

259; R&D Systems), anti-IGF-1 (ab9572), anti-follistatin (ab47941), and anti-dystrophin

(ab15277; Abcam), human laminin-α2 G4-5 (5H2; Millipore-Sigma). Rabbit anti-mouse

IgG1 secondary antibody [FITC] (NBP1-73636) was obtained from Novus Biologicals (Centennial, CO). All other secondary antibodies were obtained from Jackson ImmunoResearch.

3.4.14 Microscopy and Imaging

For all immunofluorescent images, a Zeiss Imager.Z1 with an AxioCam MRm camera, Plan-Apochromat 10x/0.45 M27 and Plan-Apochromat 20x/0.8 M27 objectives, and AxioVision40 v4.7.1.0 software was used to take single-plane multi-channel epifluorescence images. Equal exposure times were used to image corresponding channels across samples within each experiment. In all staining, DAPI and Alexa 488, CY2, or FITC, CY3, and CY5 fluorophores were used and visualized using fluorophore-appropriate filters.

3.4.15 Histological Analysis

Myofiber diameter, central nucleation, and muscle wasting analyses were performed using an ImageJ macro on 10X images of muscle sections stained with an antibody against dystrophin to outline myofibers. TUNEL(+) nuclei were counted using an ImageJ macro on 10X images of muscle sections.

3.4.16 Statistical Analysis

The statistical significance of differences between two groups was tested using an unpaired, two-tailed Student's t-test. The significance of differences between more than two groups was tested using a one-way ANOVA, with post-hoc Tukey test for multiple comparisons.

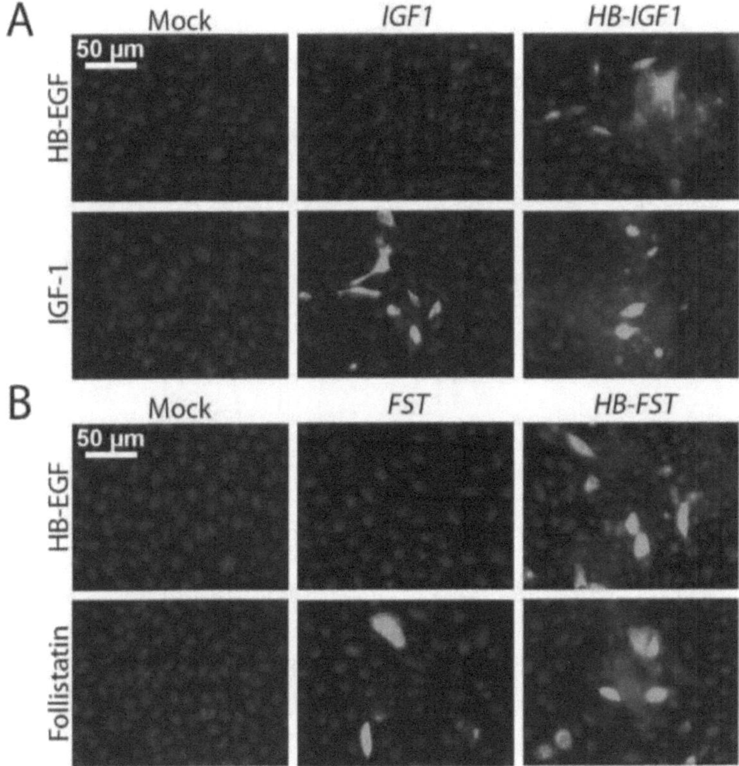

Figure 21 HB-IGF1 and HB-FST expression in transfected CHO cells

(A) CHO cells were transfected with pcDNA3.3-TOPO.CMV.*IGF1* or pcDNA3.3-

TOPO.CMV.*HB-IGF1*, and separate coverslips of cells were stained with antibodies

against HB-EGF or IGF-1 (both green). (B) CHO cells were transfected with pcDNA3.3-

TOPO.CMV.*FST* or pcDNA3.3-TOPO.CMV.*HB-FST*, and separate coverslips of cells

were stained with antibodies against HB-EGF or follistatin (both green).

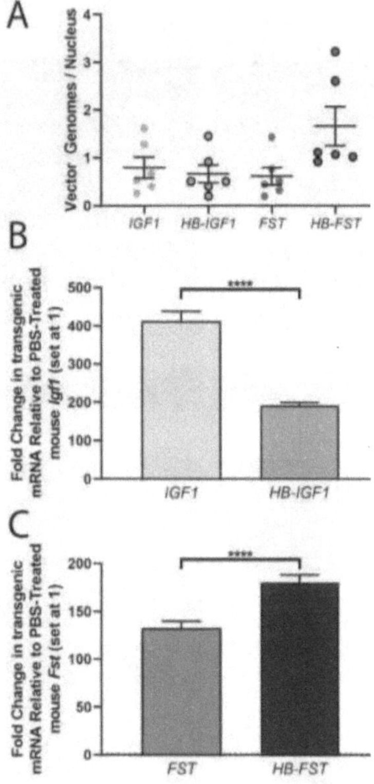

Figure 22 Biodistribution of rAAV vector genomes and transgene expression in WT mice
following IM injection

(A) Vector genome counts in the TA muscles of adult WT mice injected IM and

bilaterally with 1E+11 vg per muscle, and sacrificed for analysis 2 months post-injection.

(B) Fold change in human (transgenic) *IGF1* mRNA, including *HB-IGF1*, relative to

mouse *Igf1* in PBS-treated muscle. (C) Fold change in human (transgenic) *FST* mRNA, including *HB-FST*, relative to mouse *Fst* in PBS-treated muscle. Error bars are SEM for n

= 6 muscles per condition. ****p < 0.0001.

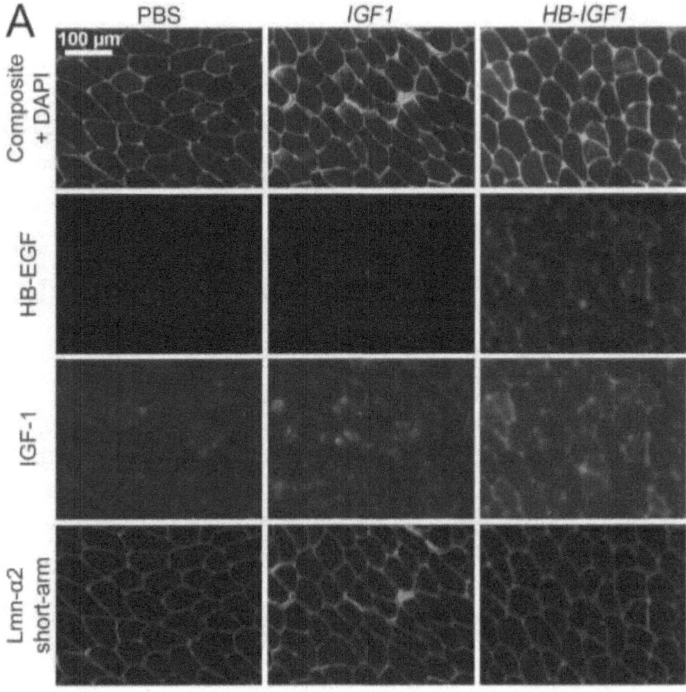

Continued.

Figure 23 Recombinant protein expression in WT mouse skeletal muscle following IM
rAAV injection

Figure 23 continued.

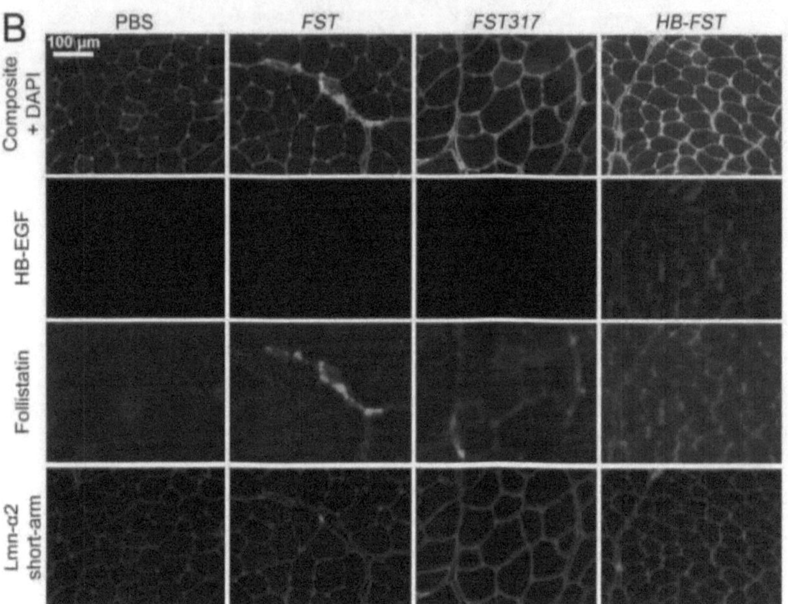

Continued.

Figure 23 continued.

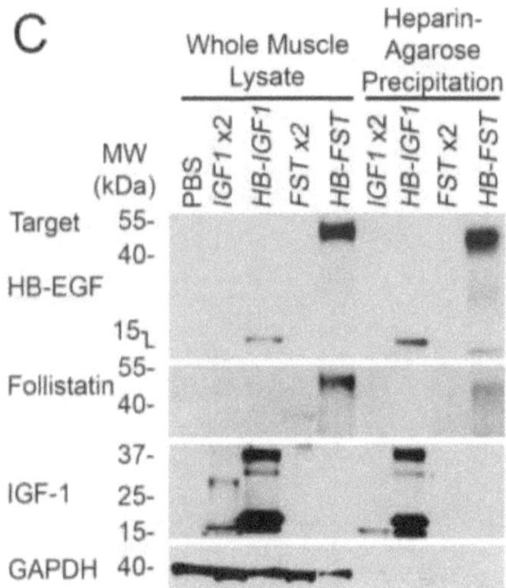

WT mouse muscle injected with PBS or rAAV9 containing trophic factor transgenes was stained 2 months post-injection with antibodies against HB-EGF (red) and either (A) IGF-1 or (B) follistatin (green) to detect recombinant protein, and counterstained with an antibody against laminin-α2 short-arm (far-red / magenta) to delineate myofibers. (C) Whole protein lysates and heparin-agarose protein precipitations from injected muscles were blotted with antibodies against HB-EGF, and follistatin or IGF-1 to detect recombinant protein, and with an antibody against GAPDH as a loading control. Twice ("2x") the microgram amount of protein in whole muscle lysate, and twice the volume of heparin-agarose precipitate was loaded for IGF1 and FST conditions compared to HB-

IGF1 and HB-FST conditions.

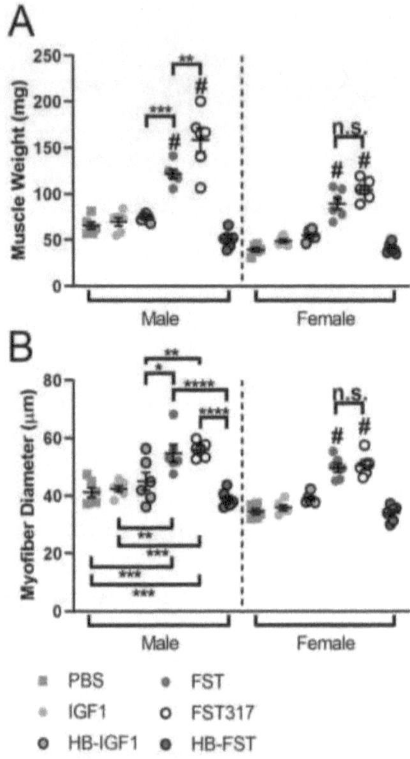

Figure 24 Muscle weight and myofiber diameters in WT mouse skeletal muscle following
IM AAV injection

PBS- or trophic factor rAAV9-injected TA muscles from wild type mice were extracted

and (A) weighed 2 months post-injection. (B) Muscle sections were stained with an

antibody against dystrophin to outline myofibers, then imaged for automated minimum

Feret's diameter calculation. Error bars are SEM for n = 6 muscles per group. *p < 0.05,

**p < 0.01, ***p < 0.001, ****p < 0.0001. # is significantly different from PBS, *IGF1*,

HB-IGF1, and *HB-FST* conditions to p < 0.0001. "n.s." is "non-significant."

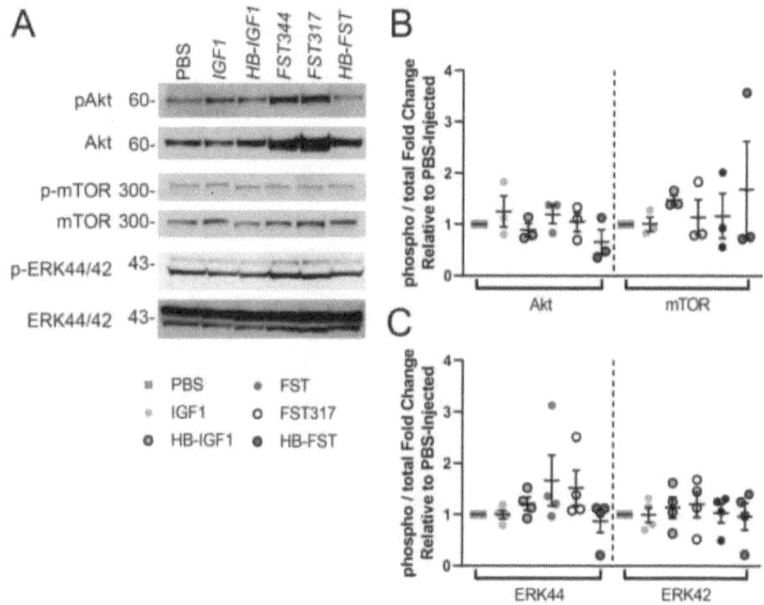

Figure 25 Phospho-protein western blots of WT mouse skeletal muscle following IM rAAV injection

(A) Lysates of WT mouse muscle injected with either PBS or rAAV were assayed for phospho-protein (p) changes relative to total protein. (B and C) Phospho-protein band density was normalized first to total protein band density, then to the ratio of phospho-protein to total protein obtained for the PBS-injected condition. Error bars are SEM for n = 3 - 4 muscles per condition.

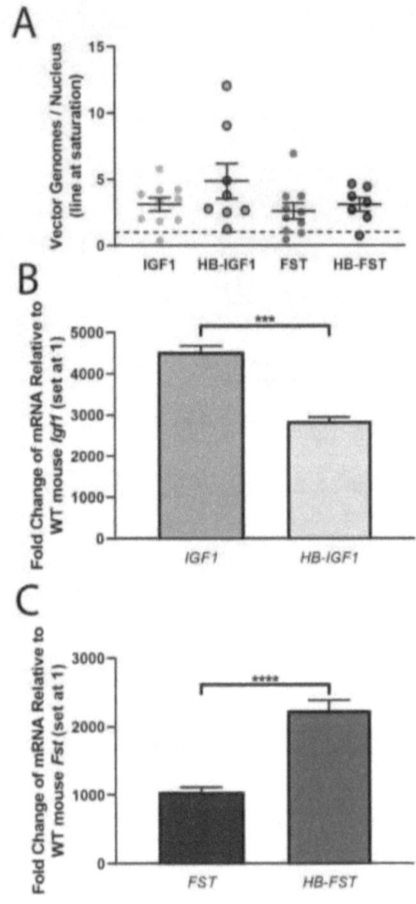

Continued.

Figure 26 Biodistribution of rAAV vector genomes and transgene expression in dy^{W} mice following IM injection

Dy^{W} mouse TA muscles were injected with 1E+11 vg trophic factor rAAV at 2 weeks

old, and euthanized for analysis 2 months post-injection. Control dy^{W} and WT littermates

Figure 26 continued.

were injected with PBS and euthanized at the corresponding time points. (A) Vg counts in the TA muscles. Error bars are SEM for n = 7 - 10 muscles per condition. (B) Fold change in human *IGF1* mRNA or HB-IGF1, relative to mouse *Igf1* in PBS-treated WT muscle. (C) Fold change in human *FST* mRNA or HB-FST, relative to mouse *Fst* in PBS-treated WT muscle. Error bars are SEM for n = 3 - 4 muscles per condition. ***p < 0.001 and ****p < 0.0001.

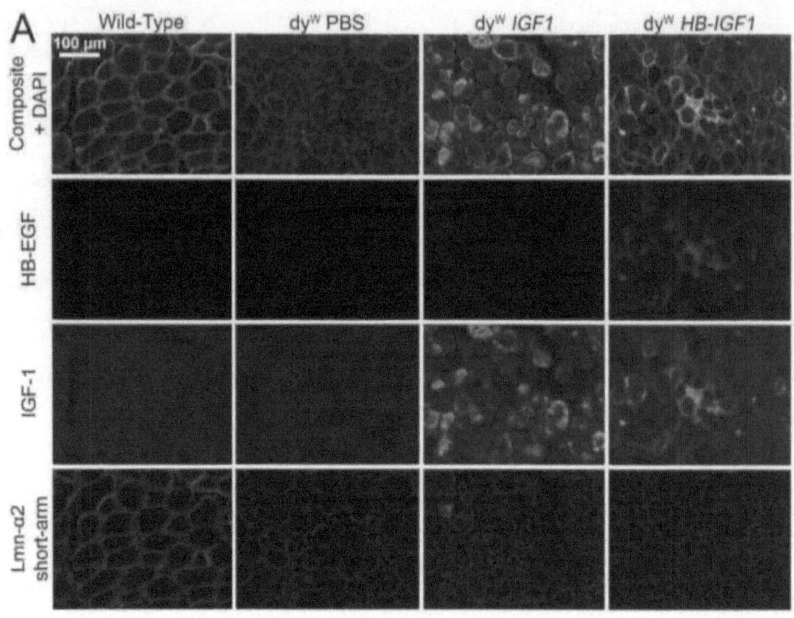

Continued.

Figure 27 Recombinant protein expression in dy^W mouse skeletal muscle following IM rAAV injection

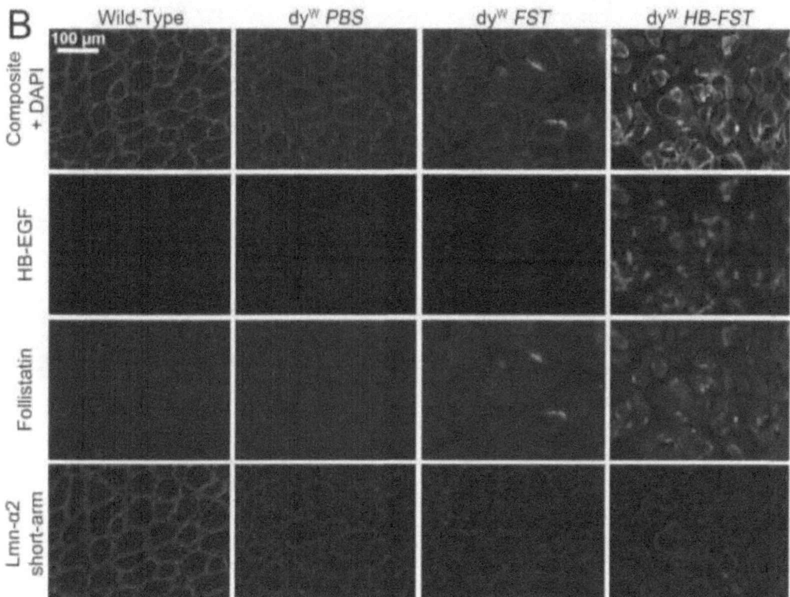

Dy^W mouse TA muscles were injected IM with (A) rAAV.CMV.*IGF1*, or .*HB-IGF1*, or (B) rAAV.CMV.*FST* or .*HB-FST* (B). At 2 months post-injection, muscle sections were stained for HB-EGF (red) and either IGF-1 (A) or follistatin (B) (green) to detect recombinant protein. Anti-laminin-α2 short-arm staining serves as both a counterstain delineating myofibers, and to distinguish dy^W muscle from WT muscle.

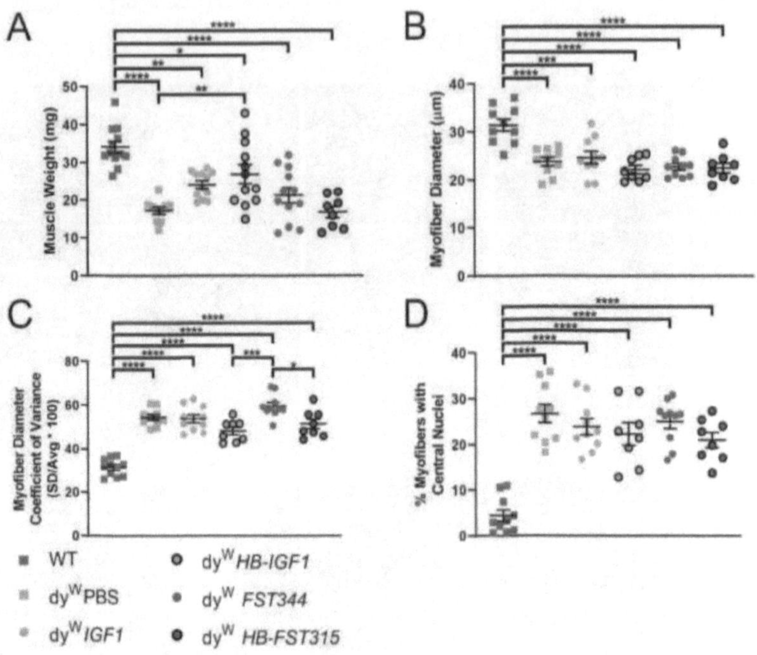

Figure 28 TA muscle weight and histopathology in dy^W mice following IM rAAV injection

Muscles were extracted from dy^W mice and WT littermates 2 months after injection with PBS or trophic factor rAAV. (A) TA muscle weight at dissection. Muscle sections were stained with an anti-dystrophin antibody to outline myofibers, then imaged for automated myofiber minimum Feret's diameter (B), diameter variation (C), and central nucleation (D) calculations. Error bars are SEM for n = 8 – 10 muscles from 4 – 5 mice, *p < 0.05, **p < 0.005, ***p < 0.0005, ****p < 0.0001.

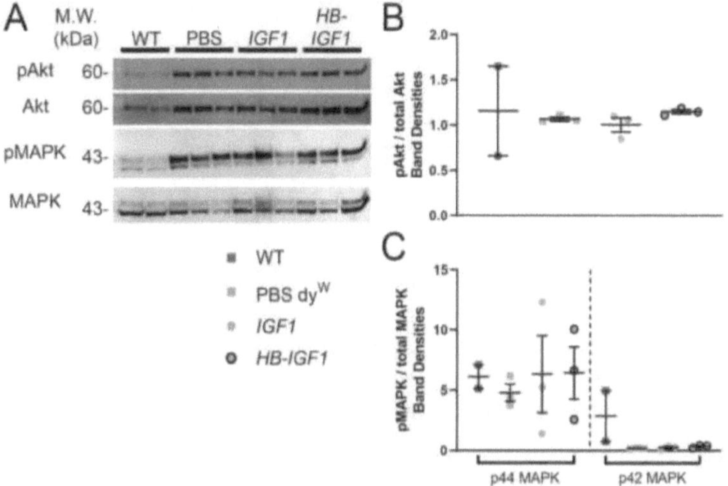

Figure 29 Akt and MAPK pathway activation in dy^W mice following IM rAAV injection Muscles were extracted from dy^W mice 2 months post-IM injection with rAAV.CMV.*IGF1*, rAAV.CMV.*HB-IGF1*, or PBS, and from age-matched WT littermates. (A) Muscle lysates were immunoblotted for phospho- (p)Akt, total Akt, pMAPK, and total MAPK. (B and C) Band densities for phosphorylated protein were quantified and normalized to band densities for total protein. Error bars are SEM for n = 2 - 3 muscles per condition.

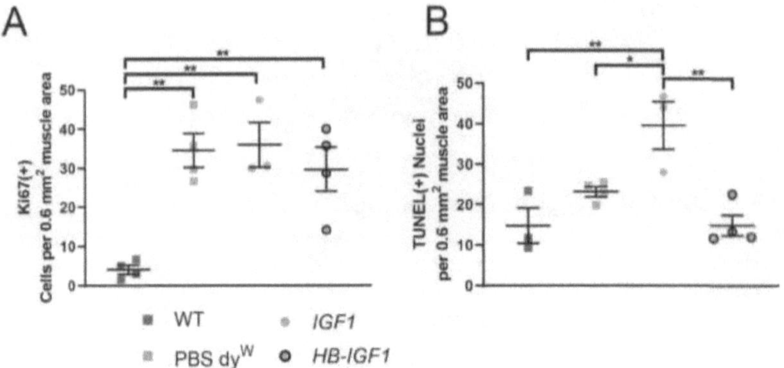

Figure 30 Cell proliferation and apoptosis in dy^W muscle following IM rAAV injection Dy^W mice were injected IM with rAAV.CMV.*IGF1*, rAAV.CMV.*HB-IGF1*, or PBS, and age-matched WT littermates were injected with PBS. Muscles were dissected 2 months post-injection. (A) Muscle sections were stained for Ki67, and Ki67-positive nuclei were counted to quantify cell proliferation per 0.6 mm² muscle area. (B) Muscle sections were TUNEL stained, and TUNEL-positive nuclei were counted to quantify apoptotic cells. Error bars are SEM for n = 3 - 4 muscles per condition. *p < 0.05, **p < 0.01.

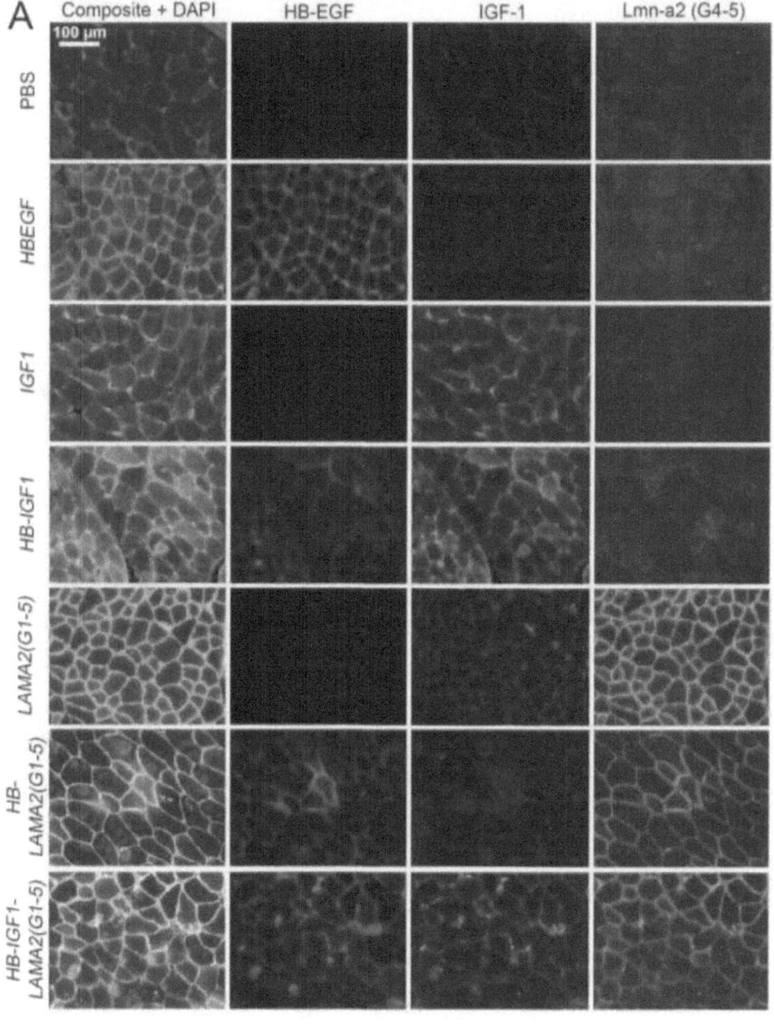

Continued.

Figure 31 HB-IGF1-LAMA2(G1-5) expression in WT muscle following IM rAAV injection

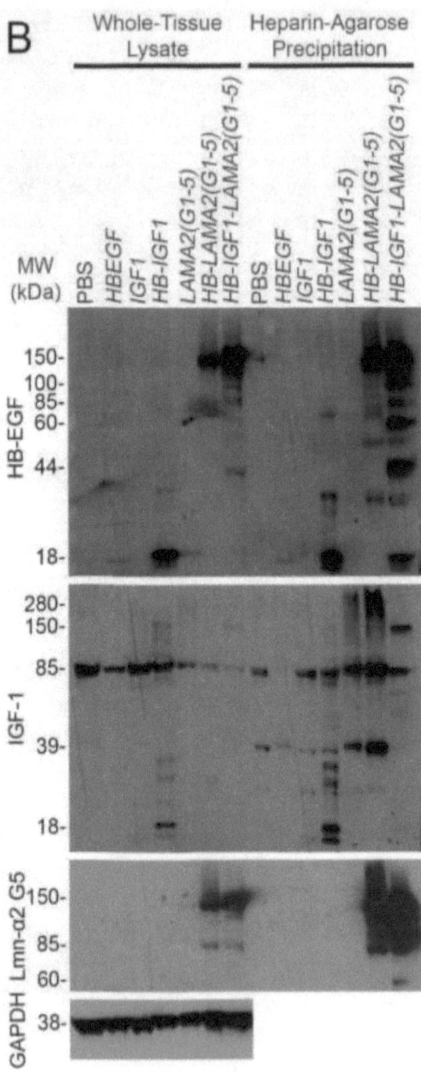

Continued.

114

Figure 31 continued.

1E+11 vg rAAV was injected IM into the TA muscles of adult WT mice. Muscles were dissected 2 months post-injection. (A) Muscle sections were stained with antibodies against human laminin-a2 G4-5 (green), HB-EGF (red), and IGF-1 (far-red / magenta). (B) Immunoblotting with antibodies against HB-EGF, IGF-1, and laminin-a2 G5 (Lmn-a2 G5) proteins, and against GAPDH as a loading control. 85 kDa and 39 kDa bands in the anti-IGF-1 blot represent non-specific secondary antibody binding.

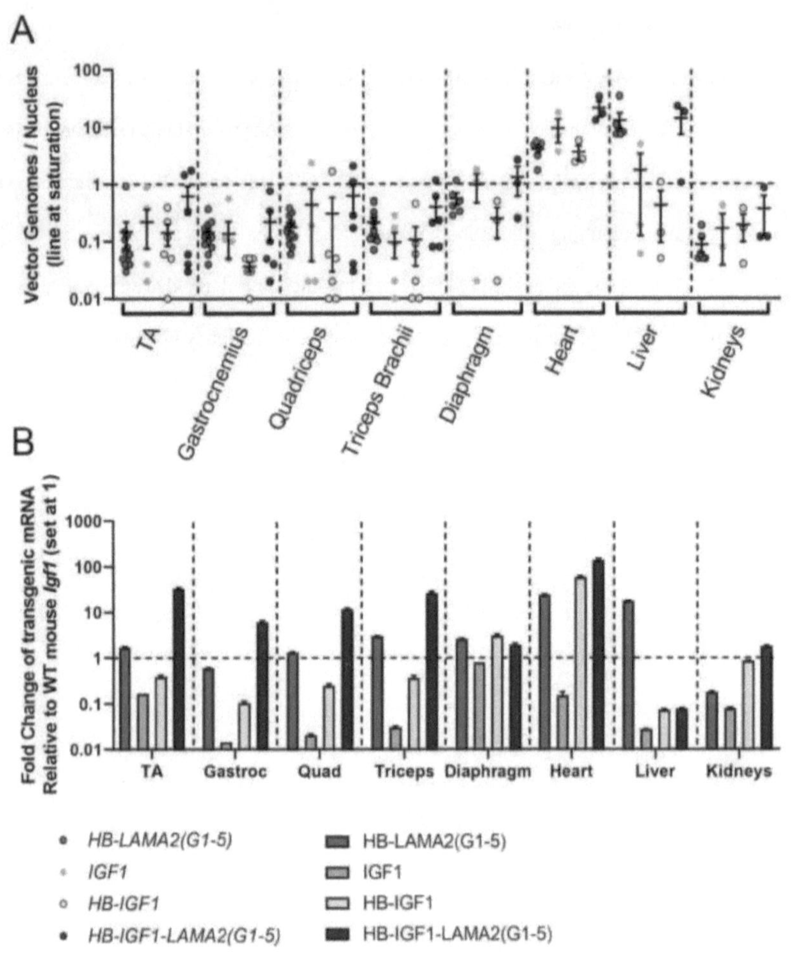

Continued.

Figure 32 Biodistribution and transgene expression in dy^W mouse tissue following neonatal IV rAAV injection

Figure 32 continued.

Neonatal dy^W mice were injected with 1E+12 vg rAAV and euthanized at 4 months of age for analysis. (A) qPCR with a standard curve of linearized rAAV genomes was used to measure biodistribution in mouse tissues. Error bars are SEM for n = 6 - 12 muscles per limb muscle and 3 - 6 samples per condition for diaphragm, heart, liver, and kidneys. (B) qRT-PCR was used to measure fold change in transgenic mRNA expression relative to endogenous mouse *Igf1* mRNA expression. Where present, error bars are SEM for n = 3 - 6 muscles per condition. Where error bars are absent, for the *IGF1* condition, indicates that only 1 muscle sample had expression.

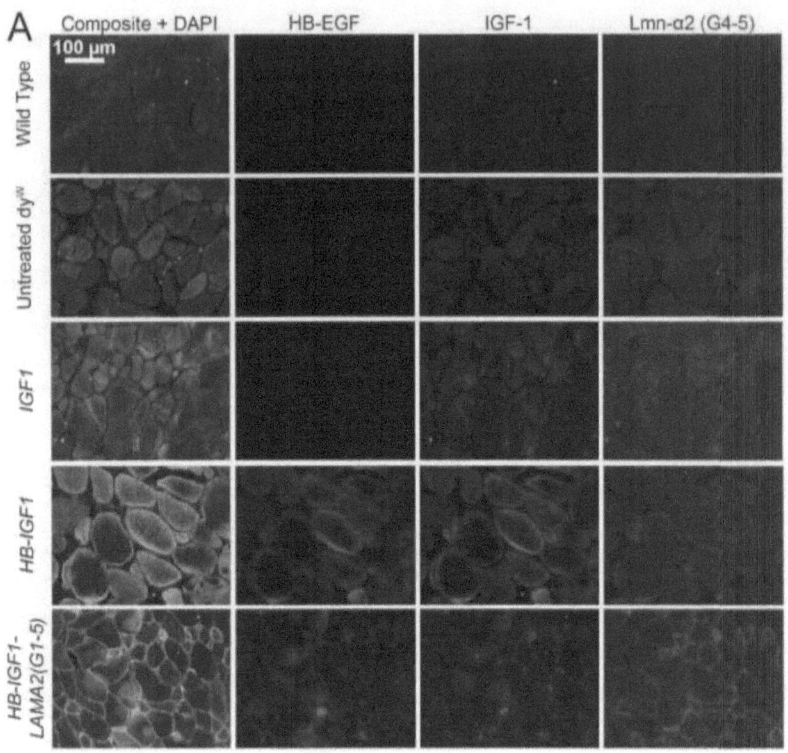

Continued.

Figure 33 Immunofluorescent staining of dy^W muscle tissue following neonatal IV rAAV injection

Figure 33 continued.

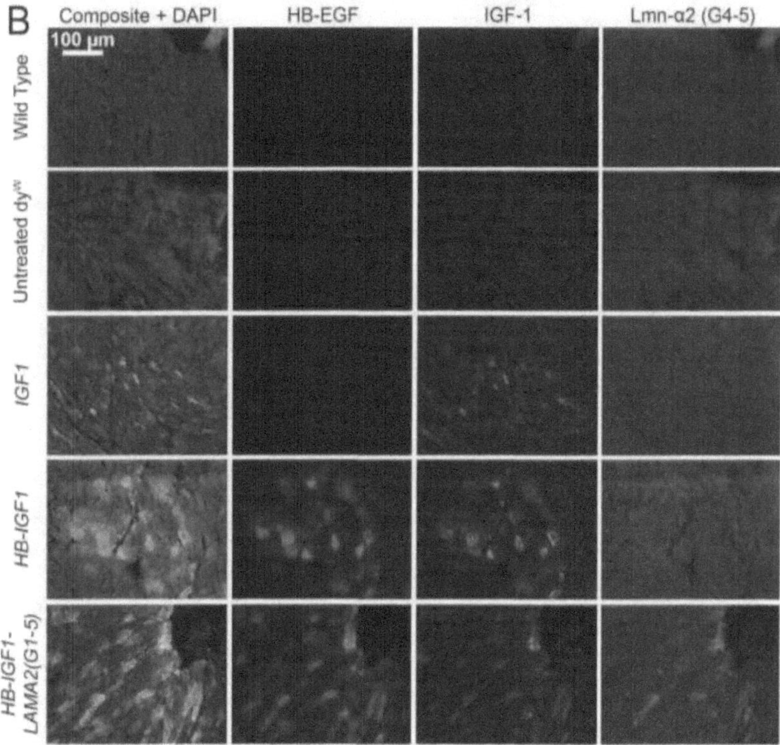

Muscle sections from the (A) triceps brachii and (B) heart were stained with antibodies against HB-EGF (red), IGF-1 (far-red / magenta), and human laminin-a2 G4-5 (green).

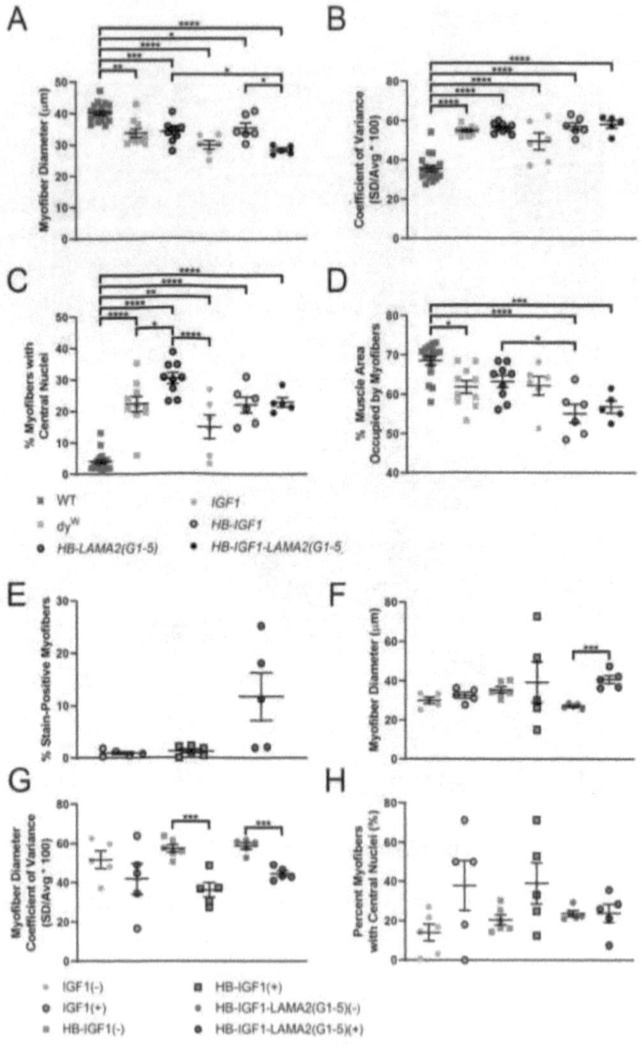

Continued.

Figure 34 Triceps brachii muscle histopathology in dy^W mice following IV rAAV

Figure 34 continued.

Triceps brachii muscle sections were stained with an anti-dystrophin antibody to outline myofibers and counterstained with DAPI to visualize nuclei. Automated image analysis was used to measure (A) average myofiber diameter, (B) myofiber diameter coefficient of variance, (C) percentage of myofibers with central nuclei, and (D) percentage of muscle area occupied by myofibers, i.e. a measurement of muscle wasting. (E) Percentage of myofibers positive for recombinant protein was determined manually. Within each condition, (F) myofiber diameters, (G) coefficients of variance, and (H) percentage of myofibers with central nuclei were stratified by recombinant protein-positive or -negative. Error bars are SEM for n = 5 - 18 muscles per condition. Significance was tested by one-way ANOVA for A - D, and by t-tests for F, G, and H. *p < 0.05, **p < 0.01, ***p < 0.001, ****p < 0.0001.

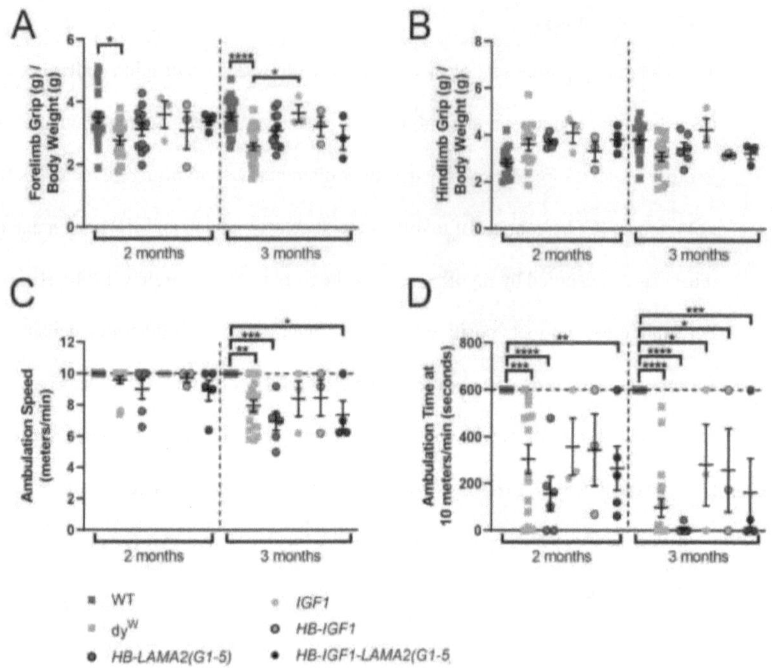

Figure 35 Grip strength and ambulation in dy^W mice following neonatal IV rAAV injection

Following IV rAAV injection in neonatal dy^W mice, grip strength and ambulation were measured at 2 months of age and at 3 months of age. (A) Body weight-normalized forelimb grip strength. (B) Body weight-normalized hindlimb grip strength. (C) Maximum speed of ambulation achieved on a treadmill, with a maximum of 10 meters / min. (D) Ambulation time at 10 meters / min on a treadmill before exhaustion, with a maximum of 600 seconds. Error bars are SEM for n = 3 - 9 mice per condition. *p < 0.05, **p < 0.01, ***p < 0.001, ****p < 0.0001.

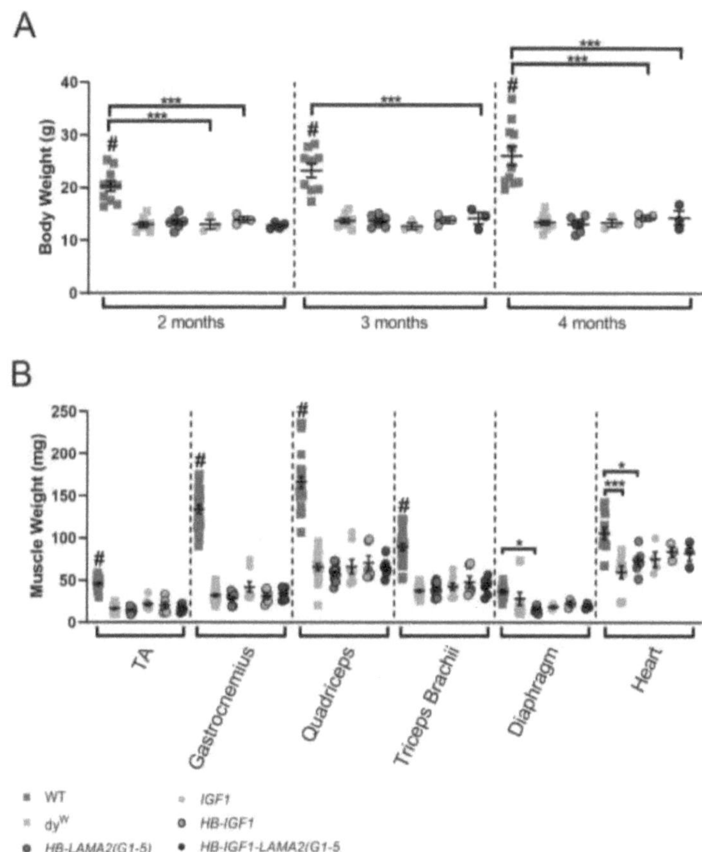

Figure 36 Body and muscle weights in dy^W mice following neonatal IV rAAV injection (A) Mice were weighed at 2, 3, and 4 months of age. (B) Muscles were weighed immediately following dissection. Error bars are SEM for n = 3 - 11 mice, 6 - 22 muscles. *p < 0.05, **p < 0.01, ****p < 0.0001. # indicates a significant difference from all other conditions to p < 0.0001, unless labeled with a bracket.

Chapter 4. Toward a Universal Gene Therapy for Dystroglycanopathies

4.1 Introduction

α-Dystroglycanopathies are severe and complex diseases caused by hypoglycosylation of α-dystroglycan. They result in a wide variety of musculoskeletal symptoms from relatively mild limb-girdle weakness and fatigability to severe muscular dystrophy with hypotonia at birth.[80, 81, 83, 287, 288] Central nervous system (CNS) abnormalities are similarly varied, from no intellectual impairment to severe brain and eye malformations.[289, 290] Several categorization schemes have been proposed, but most commonly the α-dystroglycanopathies are categorized in three overlapping groups "Walker-Warburg syndrome", "muscle-eye-brain disease" and "Fukuyama congenital muscular dystrophy".[289] Another scheme places the diseases in a more descriptive series: Congenital Muscular Dystrophy - Dystroglycanopathies (MDDG) type A - dystroglycanopathy with brain and eye anomalies, MDDG type B - dystroglycanopathy with or without mental retardation, and MDDG type C - limb-girdle.[80] Patients typically present at birth with hypotonia and arthrogryposis (joint contractures), and diagnosis can be made with a congenital muscular dystrophy genetic panel or, if the panel is inconclusive, with a muscle biopsy showing α-dystroglycan hypoglycosylation.[81, 291] There are currently no effective FDA-approved treatments for α-dystroglycanopathy that affect disease outcome.

The lengthy glycosylation sequence of α-dystroglycan begins with protein O-mannosyltransferase (POMT1/2)-mediated α-mannosylation of the serine/threonine residues of the α-dystroglycan mucin domain (Man-Ser/Thr).[292, 293] Protein O-mannose β1,4-N-acetyleglucosaminyltransferase (POMGNT2) adds an N-acetylglucosamine (GlcNAc) residue to mannose at carbohydrate 4 (GlcNAc-β1,4-Man-Ser/Thr), followed by β3-GalNAc-transferase 2 (B3GALNT2) adding an N-acetylgalactosamine (GalNAc) (GalNAc-β1,3-GlcNAc-β1,4-Man-Ser/Thr).[294] Protein O-mannose kinase (POMK) specifically phosphorylates this trisaccharide at the mannose carbohydrate 6 position.[294, 295] Then fukutin and fukutin-related protein (FKRP) sequentially add one ribitol-phosphate each to the growing polysaccharide chain,[296] followed by a xylose residue by TMEM5,[297] and a β1,4-glucuronic acid by B4GAT1 (GlcA-β1,4-Xyl-Rbo5P-Rbo5P-GalNAc-β1,3-GlcNAc-β1,4-Man-Ser/Thr).[298, 299] Like acetylglucosaminyltransferase 1 (LARGE1) then builds a variable length of glucosamine-β1,3-xylose disaccharide chain on the glucuronic acid (-[GlcA-β1,3-Xyl-α1,3]$_n$-GlcA-β1,4-Xyl-Rbo5P-Rbo5P-GalNAc-β1,3-GlcNAc-β1,4-Man(PO$_4$)-Ser/Thr).[43, 82, 300] In muscle, this heteropolysaccharide then binds laminin-211 G domains 4-5 in a Ca^{2+}-dependent manner.[162, 196, 301]

Dystroglycan contains an autoproteolytic SEA-domain between Asp556 and Cys713, allowing for the release of the extracellular α-dystroglycan domain from the transmembrane β-dystroglycan domain.[15, 17, 302] Although the C-terminal domain of α-dystroglycan subsequently binds to the β-dystroglycan N-terminal loop region, thereby anchoring α-dystroglycan to the membrane,[16, 18-22] the initial cleavage is nonetheless necessary for normal muscle development.[23] The intracellular β-dystroglycan C-terminal domain binds the C-terminus of dystrophin, connecting dystroglycan to the F-actin cytoskeleton inside the cell.[24-27] β-dystroglycan also interacts with Grb2 and SH3 adapter proteins at the membrane inner leaflet,[25, 26, 30, 32] and potentially triggers adhesion-dependent signaling pathways.[29, 33-35] Outside the plasma membrane, the extracellular α-dystroglycan protein interacts with ECM proteins including laminin, perlecan, and agrin in a glycosylation-dependent manner.[40, 43, 44]

α-Dystroglycan glycosylation is catalyzed by at least 24 unique enzymes, and loss of any one of these due to genetic mutations could cause α-dystroglycanopathy. Accordingly, there are many different mouse models of dystroglycanopathy,[89, 91, 93, 94, 98, 99, 303] however due to use-restrictions, not all are available for testing potential therapies. The Large[vls] mouse, which has a deletion in *Large* exons 3 - 5 causing a frameshift and premature stop of the *Large* gene,[102] was used in this chapter.

In order to provide direct gene replacement for each possible cause of α-dystroglycanopathy, multiple unique gene therapies would have to be developed. The objective of the work presented in this section was to develop a single gene therapy effective for α-dystroglycanopathy, regardless of the causative mutation. We sought to recapitulate the cell membrane-ECM connecting function of glycosylated α-dystroglycan in a single small "linker" protein, much as was done in Chapter 2. To do this, we fused the gene segment of *HBEGF* encoding mature form of heparin-binding epidermal growth factor-like growth factor (HB-EGF) to the 5' end of the gene segment encoding α-dystroglycan from mucin domain to the end of the C-terminal domain, creating an *HB-DAG1α* gene. The HB domain of HB-EGF has strong affinity for heparin sulfate proteoglycans[281] which are a major component of the skeletal muscle ECM.[13, 282, 283] We hypothesized that *HB-DAG1α* expression would re-establish the cell membrane-ECM connection and thereby mitigate pathology in the Large[vls] mouse model of Large-related dystroglycanopathy (MDDGA6). Since α-dystroglycanopathy involves both the

musculoskeletal and central nervous systems, we placed *HB-DAG1α* under the non-

selective cytomegalovirus (CMV) promoter, and packaged the recombinant (r)AAV

genome in a serotype 9 capsid for muscle and CNS transduction.[131, 132, 135, 136]

4.2 Results

4.2.1 HB-DAG1α expression in WT mouse muscle following IM rAAV injection

DAG1α and *HB-DAG1α* constructs were designed as described in Figure 37.

Although we were able to isolate HB-DAG1α from transfected CHO cells, as achieved

with HB-LAMA2(G1-5) and LAMA2(G1-5) in Chapter 2, staining and western blot

overlays with HB-DAG1α were unsuccessful (data not shown).

In order to determine whether recombinant HB-DAG1α and DAG1α proteins

could be successfully produced and localized in normal mouse muscle, the genes were

packaged into rAAV9 vector and 1E+11 vector genomes (vg) rAAV.CMV.*HBEGF,*

rAAV.CMV.*HB-DAG1α*, or rAAV.CMV.*DAG1α* were injected into the Tibialis Anterior

(TA) muscle of adult wild type (WT) mice (Fig. 38A). rAAV-mediated HB-EGF

overexpression resulted in marked fibrosis and accumulation of HB-EGF in the muscle

extracellular space. This was not the case for expression of HB-DAG1α protein, which

was detected at the sarcolemma with an anti-HB-EGF antibody. HB-DAG1α signal

closely localized with endogenous glycosylated α-dystroglycan, detected using a IIH6

antibody, and an antibody to dystrophin, which interacts with intracellular side of β-

dystroglycan. DAG1α protein was not detectable using an anti-HB-EGF antibody, nor distinguishable from endogenous mouse dystroglycan using anti-dystroglycan antibodies, even those made against the human protein (data not shown).

On immunoblotting with a polyclonal antibody against HB-EGF, rAAV.CMV.*HBEGF*-injected muscle lysates displayed a band at 15 kDa, while rAAV.CMV.*HB-DAG1α*-injected lysates displayed a band at 70 kDa (Fig. 38B). This weight was consistent the molecular weight of native unglycosylated αDG (72 kDa). Both bands remained following heparin-agarose precipitation of muscle lysates. Blotting with an anti-dystroglycan (α and β) antibody identified glycosylated α-dystroglycan as a large, heterogeneous protein band that was centered around 130 kDa, although the broad band was expected to be centered around 156 kDa. Endogenous β-dystroglycan was centered around 40 kDa, close to its published weight of 43 kDa. Both HB-DAG1α and DAG1α proteins appeared as relatively distinct bands of lower molecular weight than endogenous α-dystroglycan, suggesting that these recombinant proteins were less glycosylated than endogenous α-dystroglycan. This was expected, as the N-terminal domain of α-dystrolgycan is required for full native glycosylation.[79] The anti-dystroglycan band from rAAV.CMV.*HB-DAG1α*-injected muscle lysate remained following heparin-agarose precipitation and, unexpectedly, the band from the rAAV.CMV.*DAG1α*-injected muscle lysate remained as well.

4.2.2 HB-DAG1α expression in dy^W mouse muscle following IV rAAV injection.

In order to test the therapeutic efficacy of *HB-DAG1α*-expression in a mouse model of α-dystroglycanopathy, 1E+12 vg rAAV.CMV.*HB-DAG1α* were injected intravenously (IV) into the temporal vein of neonatal Large^{vls} mice. At the 4 months of age endpoint, rAAV vg biodistribution was assessed using quantitative PCR (qPCR) and a linearized vector genome standard curve (Fig. 39). As in previous IV injection experiments, vg biodistribution was generally low in limb skeletal muscles, below the saturation point of one vg per muscle nucleus typically achieved with this dose and route. Possible reasons for the low biodistribution include, suboptimal injection, a side-effect of the minute volume of Evans blue dye injected with the vector to assess injection quality, and loss of vector genome over time.

As expected from the low vg biodistribution, few myofibers were identified as expressing *HB-DAG1α* on immunofluorescent staining (Fig. 40). Myofibers that did express *HB-DAG1α* displayed an intracellular staining pattern with only some sarcolemmal pattern expected if HB-DAG1α protein bound sarcolemmal β-dystroglycan.

4.2.3 Muscle pathology in Large[vls] mice following treatment with neonatal IV rAAV.

Muscle pathology for treated and untreated Large[vls] mice was assessed, including myofiber diameter variation, and myofiber central nucleation, both indicative of muscle pathology. In order to survey a range of muscles, 2 TA, 2 gastrocnemius, 2 quadriceps, and 2 triceps brachii muscles were chosen from rAAV-treated and PBS-injected control Large[vls] mice for analysis. In limb skeletal muscles, myofiber minimum Feret's diameter was greater in WT mice (38.52 ± 0.99 μm, n = 12 triceps brachii and 7 tibialis anterior muscles from 6 mice) than in untreated (30.89 ± 1.06 μm, n = 2 tibialis anterior, 2 gastrocnemius, 2 quadriceps, and 2 triceps brachii muscles from 4 mice, p = 0.0005) and treated (31.03 ± 1.81 μm, n = 2 tibialis anterior, 2 gastrocnemius, 2 quadriceps, and 2 triceps brachii muscles from 4 mice, p = 0.0006) Large[vls] mice (Fig. 41A). Markers of pathology, myofiber diameter variation and central nuclei per myofiber (Fig. 41B and C)

were lower in WT (36.69 ± 1.37 μm/μm; 0.05 ± 0.03 CN / myofiber, n = see above) than

in untreated (54.27 ± 3.71 μm/μm; 0.68 ± 0.05 CN / myofiber, n = see above, p < 0.0001)

and treated (54.40 ± 4.33 μm/μm; 0.62 ± 0.09 CN / myofiber, n = see above, p < 0.0001)

Large^{vls} mice, however there were no significant differences in these measurements

between untreated and treated Large^{vls} mice. Interestingly, both treated and untreated

Large^{vls} mice displayed myofibers with up to six central nuclei, more than observed in

other animal models of muscular dystrophy, e.g. dy^{W} (Chapters 2 and 3).

4.2.4 Large^{vls} body and muscle weights following treatment with neonatal IV rAAV.

In general, the Large^{vls} phenotype is subtle, and body weights were not

significantly different from WT littermates at 2, 3, or 4 months of age (Fig. 42A).

Skeletal muscles and heart weights were not significantly different between Large^{vls} and

WT mice, with the exception of the TA muscle, wherein WT muscle (38.4 ± 1.2 mg, n =

12 muscles from 6 mice) weighed less than untreated (50.9 ± 2.7 mg, n = 12, p = 0.0114

muscles from 6 mice) and treated (52.2 ± 4.0 mg, n = 12 muscles from 6 mice, p = 0.005)

Large^{vls} muscles (Fig. 42B). This may be an effect of postural differences in Large^{vls}

mice, which display a perpetual slight hunch of the back, culminating in hypertrophy of

the TA muscles due to increased work and/or damage during walking. There were no

differences between treated and untreated Large^{vls} mice in body weight at any age, nor in

any other muscle weight at 4 months of age.

4.2.5 Behavior in Large^{vls} mice treated with neonatal IV rAAV.

Large^{vls} mice presented with a very mild disease phenotype. They achieved 4 months of age, the experimental endpoint, without loss of ambulation or evidence of distress, e.g. poor grooming, and unresponsiveness to stimuli. At 2 months of age, weight-normalized forelimb grip-strength was significantly greater in WT mice (4.96 ± 0.72 g/g, n = 6 mice) than in untreated Large^{vls} mice (3.82 ± 0.32 g/g, n = 6 mice, p = 0.0395), but not significantly greater than treated Large^{vls} mice (4.14 ± 0.98 g/g, n = 6 mice, p = 0.1566; Fig. 43A). There were no significant differences between conditions in forelimb grip strength at 3 months of age, nor in weight-normalized hindlimb grip-strength at 2 or 3 months of age (Fig. 43B). At 2 and 3 months of age, all WT mice were able to reach 10 meters / minute on a treadmill, and Large^{vls} mice generally were as well, with few exceptions. Ambulation time at 10 meters / minute, however, was greater in WT mice (600 seconds, without variation, n = 6 mice) than in untreated Large^{vls} mice (341 ± 177 seconds, n = 6 mice, p = 0.0347) at 2 months of age. At 3 months of age, ambulation time was greater in WT mice (600 seconds, without variation, n = 6 mice) than in both untreated Large^{vls} (244 ± 244 seconds, n = 6 mice, p = 0.0092) and treated Large^{vls} mice (272 ± 189 seconds, n = 6 mice, p = 0.0158; Fig. 43D). There were no significant differences in ambulation speed or time between treated and untreated Large^{vls} mice.

4.2.6 Muscle pathology in Large^{vls} mouse muscle following IM rAAV.CMV.*HB-DAG1α*.

To increase the rate of myofiber transduction within a single muscle, two-week-old Large^{vls} mice were injected intramuscularly (IM) in the TA with 1E+11 vg rAAV.CMV.*HB-DAG1*. Mice were euthanized and muscles were dissected two months post-injection. As expected, the transduction rate in IM injected muscles (Fig. 44A) was greater than that of muscles following IV injection (Fig. 39). There was no difference in endogenous *Dag1* mRNA expression between treated Large[vls] and untreated WT muscles (Fig. 44B). In accordance with transduction rates, a larger percentage of myofibers stained for HB-DAG1α protein in IM-injected muscle than in IV-injected muscle (Fig. 45A and E). Myofiber diameters were smaller in PBS-injected WT littermate TA muscles (31.51 ± 1.24 μm, n = 6 muscles) than in PBS-injected Large[vls] (38.98 ± 1.03 μm, n = 6 muscles, p = 0.0019) or AAV-injected Large[vls] (37.43 ± 1.42 μm, n = 6 muscles, p = 0.0110) muscles (Fig. 45B). Myofiber diameter variation, a marker of pathology, was greater in both treated (45.79 ± 0.89 μm/μm, n = 6 muscles, p < 0.0001) and PBS-injected Large[vls] muscles (44.49 ± 1.07 μm/μm, n = 6 muscles, p < 0.0001) than in PBS-treated WT muscles (31.03 ± 1.97 μm/μm, n = 6 muscles; Fig. 45C). There was no difference in myofiber diameter or diameter variation between treated and PBS-injected Large[vls] TA muscles. As seen in IV injection experiments, both rAAV-treated and PBS-injected control Large[vls] muscle had a greater average number of central nuclei per myofiber (0.52 ± 0.03 nuclei/myofiber, n = 6 muscles, p < 0.0001 and 0.60 ± 0.01 nuclei/myofiber, n = 6 muscles, p < 0.0001, respectively) than WT muscle (0.03 ± 0.01 nuclei/myofiber, n = 6 muscles; Fig. 45A and E). Notably, rAAV-treated TA muscles had significantly fewer central nuclei per myofiber than PBS-treated muscles (p = 0.0347). Within rAAV-injected muscle alone, HB-DAG1α-positive myofibers had significantly fewer central nuclei per myofiber (0.26 ± 0.04 nuclei/myofiber, n = 6 muscles) than HB-

DAG1a-negative myofibers (0.58 ± 0.02 nuclei/myofiber, n = 6 muscles, p < 0.0001; Fig. 45A and E).

4.3 Discussion

α-Dystroglycanopathy may be caused by mutations in any one of the 24 or more genes responsible for the complete glycosylation of α-dystroglycan. Without complete glycosylation, interaction between dystroglycan and ECM elements, most notably laminins, is reduced, thereby reducing overall cell membrane-ECM connection. This results in mild to severe muscular dystrophy with a range of extramuscular involvement including defects in central nervous system development.[80, 81, 83, 287-290] Since mutations in multiple genes may cause α-dystroglycanopathy, gene replacement or gene correction therapies would have to address each gene individually to cover the entire patient population. In this chapter, we took a functional replacement approach, rather than gene replacement or correction, to address the main pathogenic issue in dystroglycanopathy: the cell membrane-ECM connection. To do this, we created a linker protein comprised of an ECM-binding domain, and the β-dystroglycan-binding domain of α-dystroglycan. This linker protein would, theoretically, restore the cell membrane-ECM connection missing in dystroglycanopathy.

Large[vls] mice proved suboptimal as an animal model of dystroglycanopathy due to a very mild phenotype, even though dystroglycanopathy presents in a wide range of severity in humans. Behavior (Fig. 41), and body and muscle weights (Fig. 40), with the exceptions of ambulation time and TA muscle weight, were not significantly different from those of WT littermates. On the other hand, muscle pathology was evident in the form of reduced myofiber diameters (except in the TA muscle), increased diameter variation, and marked central nucleation relative to WT muscle (Figs. 5 and 9). In this sense, Large[vls] mice are similar to previously studied Large[myd] mice.[99]

Low transduction rates in mouse tissue undermined the assessment of rAAV.CMV.*HB-DAG1α* treatment efficacy with IV delivery. It was therefore not possible to determine whether the lack of improvement in rAAV9.CMV.*HB-DAG1α*-treated mice was due to a failure of treatment efficacy, or of treatment delivery. IM injections improved the transduction and expression rates. While other markers of muscle pathology did not improve in IM-treated mice, there was a significant decrease in central nucleation. This effect was due almost exclusively to the HB-DAG1α-positive myofibers which had fewer central nuclei per myofiber than negative myofibers. In rodent muscular dystrophy models, the presence of central nuclei in mature mice is interpreted as a reflection of muscle degeneration-regeneration cycles.[74, 99, 304, 305] Pertinent to dystroglycanopathy, β-dystroglycan can migrate to the nucleus in certain cells, including muscle cells.[38] At the nucleus, β-dystroglycan has roles in regulating nuclear functions, including nuclear morphology and interaction with centrosomes.[37, 39] Furthermore, β-dystroglycan interacts with nuclear lamins known to be involved in the anchoring of muscle nuclei at the neuromuscular junction.[306] It may be that anchoring β-dystroglycan to the ECM via α-dystroglycan lowers the amount of β-dystroglycan translocating to the nucleus, and thereby affects the location of the nucleus. If so, then reduced β-dystroglycan anchoring by hypoglycosylation of α-dystroglycan in Large[vls] mice may also affect nuclear location. Restoring β-dystroglycan anchoring with HB-DAG1α-expression, if this is indeed the case, may restore peripheral nuclear location in muscle and explain the observed results.

These studies represent a starting point for a universal gene therapy for α-dystroglycanopathy, targeting the sarcolemma-ECM connection, rather than a specific gene. Both the abundance of central nuclei and reduction in central nuclei with HB-DAGα expression indicate a role for dystroglycan-ECM binding in nuclear localization to be investigated further. For future studies, IV delivery of rAAV9.CMV.*HB-DAG1* must be improved and re-assessed for therapeutic efficacy. Ideally, a dystroglycanopathy mouse model with a more severe phenotype would be studied. As glycosylation may not be required for this method to work, studies to shorten HB-DAG1 to contain only the C-terminal domain might also be tried to make room for adding other structural elements. Of course, it may also be that native glycosylation is required for dystroglycan function in all circumstances, and that the partial changes seen here reflect the need for the natively glycosylated protein for full function.

4.4 Materials and Methods

4.4.1 Production of AAV9 vectors containing *DAG1α* and *HB-DAG1α* genes

The *DAG1α* and *HB-DAG1α* genes were designed using the human *DAG1* gene segment encoding the mucin domain through the α-dystroglycan C-terminal domain from NM_001165928.3 (base pairs 1716 to 2738), the human HBEGF signal peptide from NM_001945.2 (base pairs 276 to 348) for *DAG1α*, and the human HBEGF signal peptide through heparin-binding domain from NM_001945.2 (base pairs 276 to 597) for *HB-DAG1α*. The recombinant genes were synthesized by GeneArt Gene Synthesis (Regensburg, Germany) in a pMA-T vector backbone. The genes were subcloned into a pcDNA3.3-TOPO backbone for use in cell culture transfections, and into a rAAV plasmid with CMV promoter, SV40 enhancer, and SV40 poly-adenylation signal for rAAV vector production. From 5' ITR to 3' ITR, the genome length of rAAV.CMV.*DAG1α* was 2.502 kb, and the length of rAAV.CMV.*HB-DAG1α* was 2.751 kb. Plasmids were packaged in rAAV9 vectors by the Nationwide Children's Hospital Viral Vector Core (Columbus, OH) using the triple transfection technique in HEK293 cells,[130] and highly purified using sucrose density centrifugation and anion exchange chromatography, as previously described.[202]

4.4.2 Cell Transfection

CHO-K1 cells (ATCC, Manassas, Virginia) were grown in 10 cm dishes in DMEM (Thermo Fisher Scientific) with 10% fetal bovine serum (Millipore-Sigma). When at 80% confluence, cells were transfected with pcDNA3.3-TOPO.CMV.*DAG1α* or pcDNA3.3-TOPO.CMV.*HB-DAG1α* using TransIT-X2 Dynamic Delivery System (Mirus, Madison, Wisconsin), according to the manufacturer's instructions for CHO-K1 cells in 10 cm dishes.

4.4.3 Mice

All mouse experiments were performed using protocols approved by the Institutional Animal Care and Use Committee at The Abigail Wexner Research Institute at

Nationwide Children's Hospital (Columbus, OH). C57BL/6J mice and Large[vls] mice[102] were purchased from Jackson Laboratories (Bar Harbor, ME).

4.4.4 Intramuscular AAV Injections

Adult mice were anesthetized with isofluorane (Baxter, Deerfield, IL), and hindlimbs were shaved over the TA muscles. 1E+11 vg rAAV9 in 25 µL PBS was injected into each TA muscle using a 0.33cc Thinpro Insulin Syringe (Terumo Medical, Somerset, NJ). Two-week-old mice were restrained and 1E+11 vg rAAV in 5 µL PBS was injected into each TA muscle using a 0.3cc Thinpro Insulin Syringe.

4.4.5 Intravenous AAV Injections

Mouse pups up to 48 hours old were anesthetized on ice and injected with 1E+12 vg rAAV9 into the temporal vein using a Thinpro 0.3cc Insulin Syringe. The minimum injection volume allowable by virus titer was used.

4.4.6 Grip Strength

On three consecutive days, one week prior to grip strength assessment, mice were trained on the Grip Strength Meter (Columbus Instruments, Columbus, OH). Each training day consisted of five forelimb and five hindlimb grip strength trials which were not recorded. The week of assessment, grip strength was assessed on five consecutive days. Each assessment consisted of five forelimb trials using the "T Peak" setting and five hindlimb trials using the "C Peak" setting on the Grip Strength Meter and measured in grams, followed by body weight measurement. For each mouse, on each day of the assessment week, the five trials were averaged and normalized to body weight obtained on that day.

For each mouse, the final grip strength value was the average of the body weight-normalized grip strength values obtained over the five days.

4.4.7 Ambulation

On three consecutive days, one week prior to ambulation assessment, mice were trained on the Treadmill Meter (Columbus Instruments, Columbus, OH). The week of assessment, mice were placed on the stationary Treadmill Meter, one or two mice per enclosed lane. The treadmill was started at 5 meters per minute, increasing 1 meter per minute every 30 seconds until 10 meters per minute was reached. The treadmill was run at 10 meters per minute for 10 minutes. For each mouse, the trial ended when the mouse fell off the end of the treadmill three times in a span of 20 seconds. At the end of the trial, the maximum speed reached and time of ambulation at 10 meters per minute were recorded. If a mouse failed to reach 10 meters per minute, 0 seconds was recorded for ambulation time. For each mouse, the final values for speed and ambulation time were the averages of those obtained over the five days.

4.4.8 Tissue Processing

Skeletal muscles were removed, weighed, mounted with O.C.T. (Thermo Fisher Scientific) on cork, and snap-frozen liquid nitrogen-cooled 2-methylbutane. The internal organs and heart were removed, placed in Peel-A-Way Disposable Embedding Molds (Thermo Fisher Scientific), covered in Tissue Plus O.C.T. (Thermo Fisher Scientific) and frozen in a 2-methylbutane (Thermo Fisher Scientific) / dry ice bath.

4.4.9 AAV Biodistribution

DNA was extracted from tissue shavings using a DNeasy Blood & Tissue Kit (Qiagen, Germantown, MD). Vector genome quantity in 100 ng of sample DNA was determined using 2X TaqMan Gene Expression Master Mix, a standard curve of linearized rAAV plasmid from 5 vg to 5E+6 vg, and the following primer/probe set amplifying a region within the CMV promoter: 5'-CTAACGCCAATAGGGACTTTCC-3' (forward), 5'- 56-FAM/ACGGTAAAC/ZEN/TGCCCACTTGGCAGTACA/3IABkFQ/-3' (probe), and 5'-GGGCGTACTTGGCATATGATAC-3' (reverse). Vector genomes per mouse nucleus were then calculated using an estimate of 1.7E+5 mouse nuclei per 1 μg of genomic DNA.

4.4.10 Gene Expression Measurement

RNA was extracted from tissue shavings using a Direct-zol RNA MiniPrep kit (Zymo Research, Irvine, CA), and cDNA was prepared using a High-Capacity cDNA Reverse Transcriptase kit (Applied Biosystems, Bedford, MA). Transgene expression was measured in cDNA samples using TaqMan Gene Expression Master Mix and the following probe/primer set amplifying a region overlapping the HB domain and DAGα domain: 5'-GCACTGGCCACACCAAACAA-3' (forward), 5'- 56-FAM/GGCTAGGGA/ZEN/AGAAGAGGGACCCACA/3IABkFQ/-3' (probe), and 5'-ATCCATGCTACACCCACACCTG-3' (reverse). Additional assays included mouse *Dag1* Assay Mm00802400_m1 (Applied Biosystems). 18S rRNA was used as an internal control (Applied Biosystems). The $2^{-\Delta\Delta CT}$ method was used to calculate fold-change in expression.[286]

4.4.11 Heparin-Agarose Precipitation

200 µL Heparin-Agarose Type I, saline suspension (Millipore-Sigma) was added per mL of conditioned media, cell lysate, or muscle lysate to be precipitated, and mixed overnight at 4°C. Precipitated samples were washed thoroughly with PBS. Precipitated samples were resuspended in 50 µL 1X NuPage LDS Sample Buffer with 100 mM β-mercaptoethanol. 20 µL of the resuspension was boiled and used for immunoblotting.

4.4.12 Immunoblotting

Protein was extracted from tissue shavings using NP40 lysis buffer: : 1% (w/v) NP40 (Millipore-Sigma, St. Louis, MO) in 50 mM Tris buffer with 150 mM NaCl, 1 mM EDTA (Invitrogen, Grand Island, NY), and cOmplete EDTA-free Protease Inhibitor Cocktail (Millipore-Sigma). 40 µg of each protein lysate was added to NuPAGE LDS Sample Buffer (Thermo Fisher Scientific) with 100 mM β-mercaptoethanol (Thermo Fisher Scientific) and boiled for 10 minutes prior to gel electrophoresis and transfer to nitrocellulose membranes for immunoblotting. Antibodies included anti-HB-EGF (AF-259) and anti-dystroglycan (AF-6868; R&D Systems, Minneapolis, MN), and anti-GAPDH (G9545; Millipore-Sigma, Burlington, MA). The HRP-conjugated secondary antibodies against goat and sheep IgG were obtained from Jackson ImmunoResearch (West Grove, PA). Protein band densities were measured using ImageJ 1.46r.

4.4.13 Immunofluorescent Staining

10 µm tissue sections, cut on a Cryostat, were placed on Fisher Superfrost Plus glass slides (Thermo Fisher Scientific). A hydrophobic border was drawn around each section using an ImmEdge Hydrophobic Barrier PAP pen (Vector Labs, Burlingame, CA). Sections were fixed in 2% Paraformaldehyde (EMS) in PBS for 10 minutes,

permeabilized in 0.1% Triton X-100 (Millipore-Sigma) in PBS for 10 minutes, incubated

in Mouse-on-Mouse Blocking Reagent (Vector Labs) for 1 hour, and blocked with 5%

donkey serum in PBS for 30 minutes prior to incubation overnight at 4°C in primary

antibodies. Sections were then incubated with secondary antibodies for 1 hour at room

temperature and then treated with ProLong Gold with DAPI (Invitrogen) before adding a

coverslip. Antibodies included anti-HB-EGF (AF-259; R&D Systems), anti-α-

dystroglycan (IIH6, 05-593, Millipore-Sigma) and anti-dystrophin (ab15277; Abcam).

All secondary antibodies were obtained from Jackson ImmunoResearch.

4.4.14 Microscopy and Imaging

For all immunofluorescent images, a Zeiss Imager.Z1 with an AxioCam MRm camera,

Plan-Apochromat 10x/0.45 M27 and Plan-Apochromat 20x/0.8 M27 objectives, and

AxioVision40 v4.7.1.0 software was used to take single-plane multi-channel

epifluorescence images. Equal exposure times were used to image corresponding

channels across samples within each experiment. In all staining, DAPI and Alexa 488,

CY2, or FITC, CY3, and CY5 fluorophores were used and visualized using fluorophore-

appropriate filters.

4.4.15 Histological Analysis

Myofiber diameter, central nucleation, and muscle wasting analyses were performed

using an ImageJ macro on 10X images of muscle sections stained with an antibody

against dystrophin to outline myofibers.

4.4.16 Statistical Analysis

The significance of differences between more than two groups was tested using a one-way ANOVA, with post-hoc Tukey test for multiple comparisons.

4.5 Figures

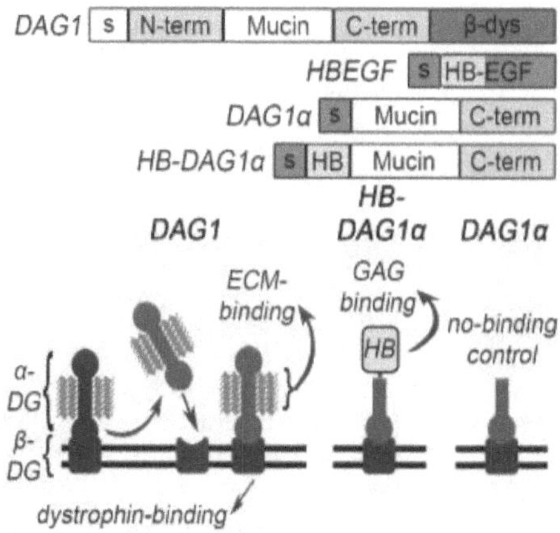

Continued.

Figure 37 Design of the HB-DAG1α gene

The *HB-DAG1α* gene was designed by fusing the segment of *HBEGF* gene encoding the signal peptide through the end of the heparin-binding (HB) domain to the 5' end of the *DAG1* gene segment encoding the α-dystroglycan mucin domain through the end of the α-dystroglycan C-terminus. The DAG1α C-terminal domain is expected to bind free β-dystroglycan transmembrane protein (β-DG) following cleavage of the endogenous α-dystroglycan domain (α-DG), while the HB domain is expected to bind glycosaminoglycans (GAGs) in the ECM. The N-terminal domain is normally cleaved off of the protein during intracellular processing and is not present on the mature protein. The intracellular C-terminus of β-dystroglycan binds dystrophin, thereby connecting to

Figure 37 continued.

the cytoskeletal network in muscle. The *DAG1α* gene was designed as a negative control

treatment condition encoding the signal peptide of HB-EGF fused to the mucin domain

through C-terminus of α-dystroglycan. Lacking a matrix-binding domain, secreted

DAG1α protein is expected to bind β-dystroglycan but fail to establish cell-ECM

connection.

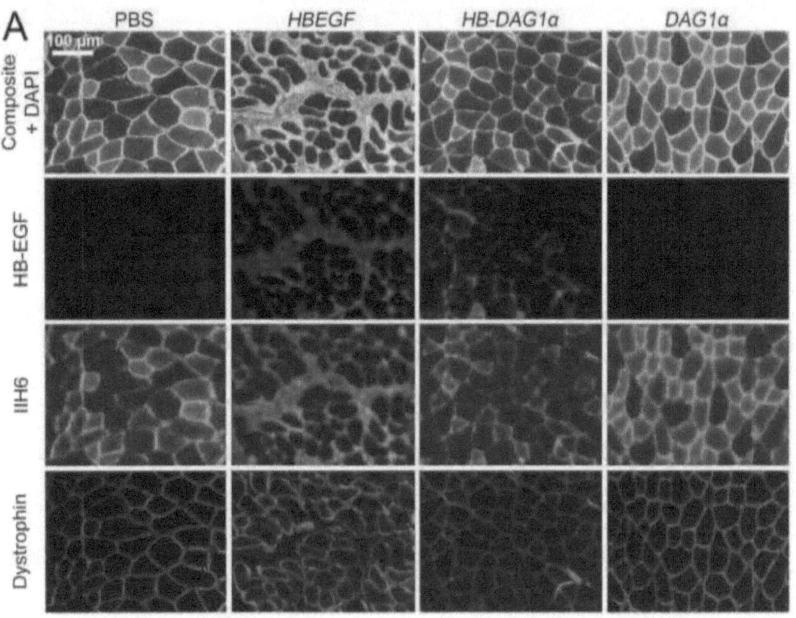

Continued.

Figure 38 HB-DAG1α protein expression in WT mouse skeletal muscle following
intramuscular rAAV injection

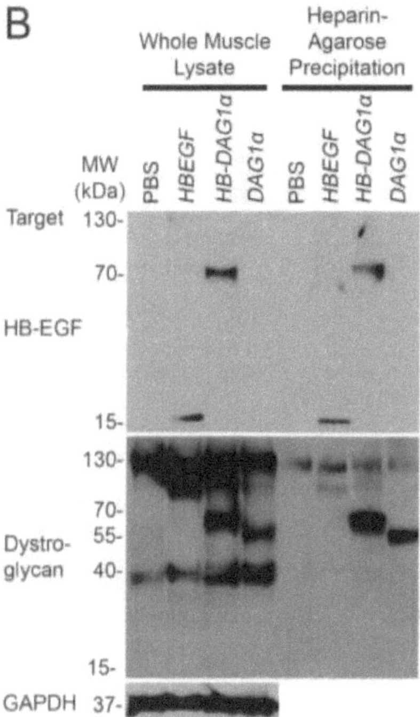

TA muscles of adult WT mice were injected with 1E+11 vg rAAV.CMV.*HB-DAG1α*, rAAV.CMV.*DAG1α*, rAAV.CMV.*HBEGF* or an equivalent volume of PBS. Muscles were analyzed 2 months post-injection. (A) TA muscles were stained with antibodies against HB-EGF (red), IIH6 (green), and dystrophin (far red / magenta). (B) Recombinant human HB-EGF protein (rHB-EGF), and TA muscle lysates were immunoblotted with antibodies against HB-EGF and dystroglycan. Anti-GAPDH immunoblotting was performed as a loading control. A heparin-agarose precipitation of muscle lysates was performed to demonstrate heparin-binding activity in proteins.

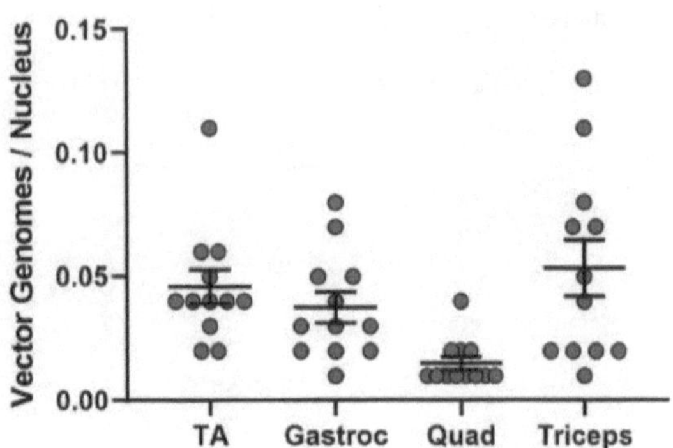

Figure 39 Biodistribution of vector genomes in Large[vls] mice following neonatal intravenous rAAV injection

P1 Large[vls] mice were injected intravenously with 1E+12 vg rAAV.CMV.*HB-DAG1a* and euthanized four months post-injection. The number of rAAV vector genomes per nucleus was quantified by qPCR using a linearized vector genome standard. Error bars are standard error of the mean (SEM) for n = 12 muscles from 6 mice per bar.

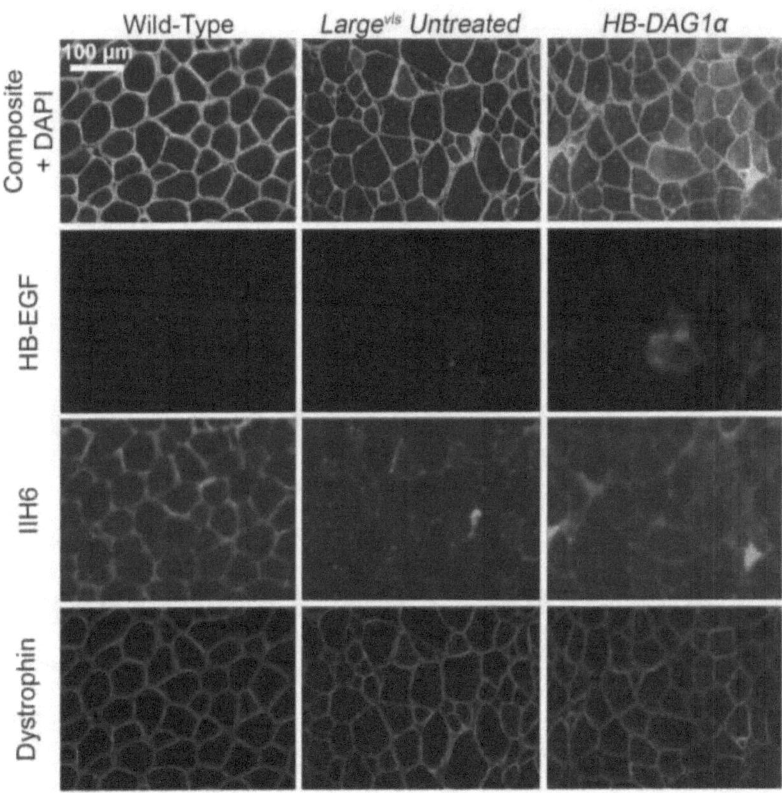

Figure 40 HB-DAG1α protein expression in Large^vls mouse triceps brachii muscle following neonatal IV rAAV injection

Muscle sections from IV rAAV.CMV.*HB-DAG1α*-injected Large^vls mice were stained with antibodies against HB-EGF (red), α-dystroglycan (IIH6, green), and dystrophin (far-red / magenta).

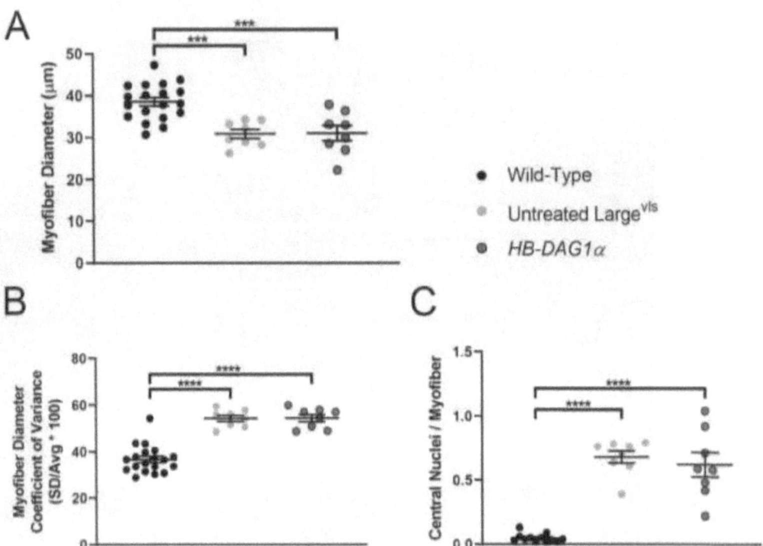

Figure 41 Muscle pathology in Large^{vls} mice following neonatal intravenous rAAV injection

P1 Large^{vls} mice were injected IV with 1E+12 vg rAAV.CMV.*HB-DAG1α* and euthanized four months post-injection. (A) Minimum Feret's diameter measurements of myofibers in limb skeletal muscles. (B) Myofiber diameter coefficient of variance in limb skeletal muscles. (C) Average number of central nuclei per myofiber in limb skeletal muscles. Error bars are SEM for n = 12 triceps brachii and 7 tibialis anterior (TA) muscles for WT; n = 2 TA, 2 gastrocnemius, 2 quadriceps, and 2 triceps brachii muscles from 4 mice for untreated and *HB-DAG1a*-treated Large^{vls} mice. ***p < 0.001, ****p < 0.0001.

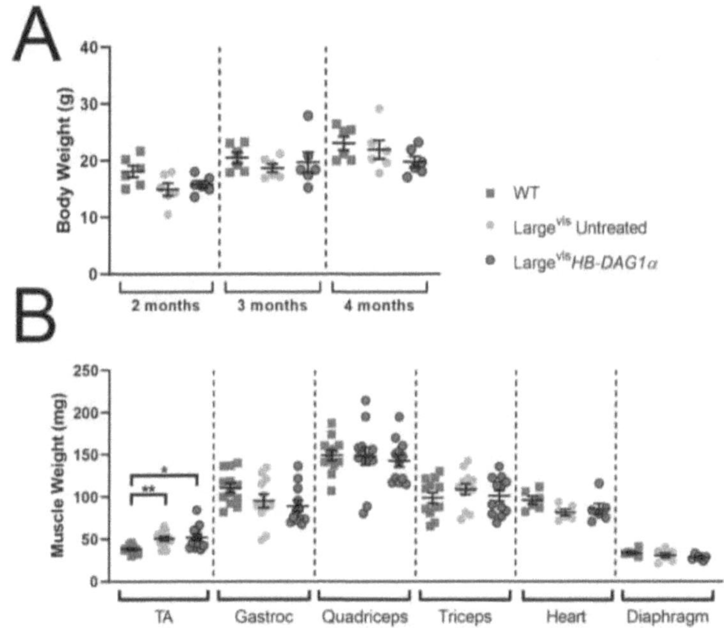

Figure 42 Body and muscle weights in Large^{vls} mice following neonatal intravenous rAAV injection

P1 Large^{vls} mice were injected IV with 1E+12 vg rAAV.CMV.*HB-DAG1α* and euthanized four months post-injection for dissection and analysis. (A) Body weight was measured at 2, 3, and 4 months of age. (B) Muscle weights were measured immediately post-dissection at 4 months of age. Error bars are SEM for n = 6 mice, 12 of each limb muscle, 6 diaphragms, and 6 hearts. *p < 0.05, **p < 0.01.

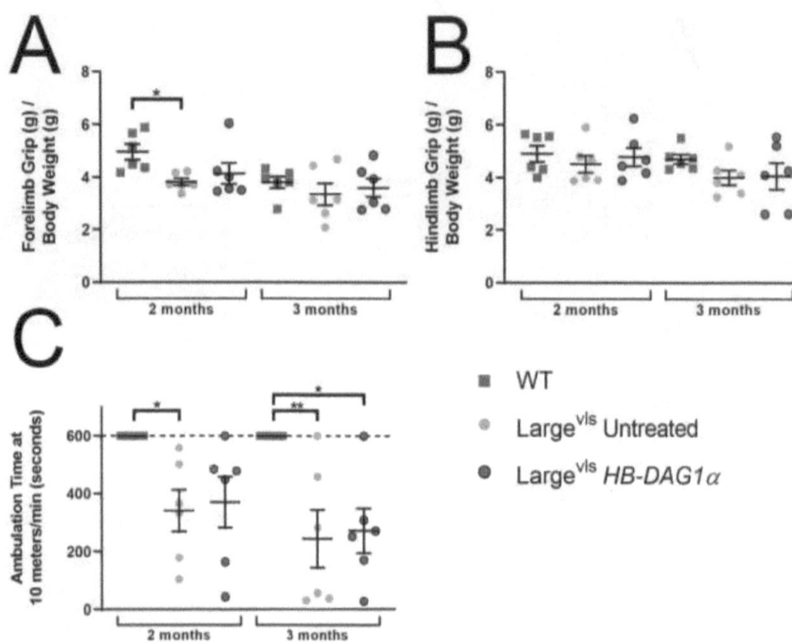

Figure 43 Behavioral measurements in Large^{vls} mice following neonatal intravenous rAAV injection

P1 Large^{vls} mice were injected intravenously with 1E+12 vg rAAV.CMV.*HB-DAG1α*, and behavior was measured at 2 and 3 months of age. (A) Body weight-normalized forelimb grip-strength. (B) Body weight-normalized hindlimb grip-strength. (C) Ambulation time at 10 meters / minute, arbitrarily capped at 600 seconds. Error bars are SEM for n = 6 mice. *p < 0.05, **p < 0.01.

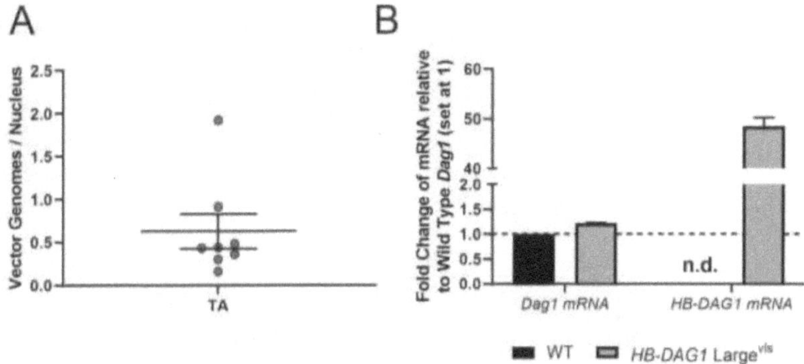

Figure 44 Biodistribution of vector genomes and transgene expression in Large[vls] mice following juvenile intramuscular rAAV injection

The TA muscles of two-week old Large[vls] mice were injected with 1E+11 vg

rAAV.CMV.*HB-DAG1α* or PBS and harvested for analysis at 2 months post-injection.

(A) The number of rAAV vector genomes per nucleus was quantified by qPCR using a

linearized vector genome standard. (B) *HB-DAG1α* mRNA expression was quantified by

qRT-PCR and fold-change was calculated relative endogenous mouse *Dag1* mRNA

expression in age-matched, untreated WT TA muscle. Error bars are SEM for n = 8

muscles from 4 mice. "n.d." = not detected.

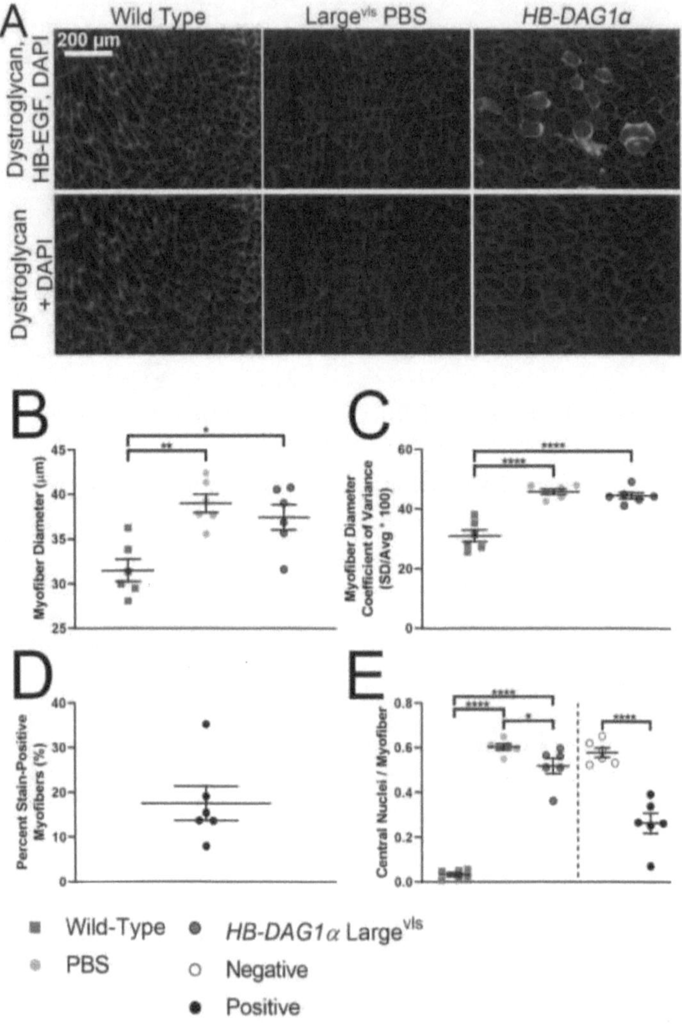

Continued.

Figure 45 Muscle pathology in Large^vls mice following juvenile intramuscular rAAV9 injection

Figure 45 continued.

The injected TA muscles of juvenile Large^{vls} mice were stained and imaged for semi-automated image analysis. (A) Examples of immunofluorescent images used for image analysis. Dystroglycan (red) was used to outline myofibers. HB-EGF staining (green) was used to identify HB-DAG1a-expressing myofibers. DAPI (blue) was used to identify nuclei and assess myofiber central nucleation. (B) Minimum Feret's diameter of myofibers. (C) Myofiber diameter coefficient of variance. (D) Percent of myofibers staining positively for HB-DAG1α protein using the HB-EGF antibody. (F) Number of central nuclei counted per myofiber counted, including per myofiber in rAAV.CMV.*HB-DAG1a*-treated muscle (*HB-DAG1α* Large^{vls}), and stratified by expressing and non-expressing, i.e. per HB-DAG1α-expressing myofiber in treated muscle (positive) and per non-expressing myofiber in treated muscle (negative) muscle. Error bars are SEM for n = 6 muscles from 3 mice per condition. *p < 0.05, **p < 0.01, ****p < 0.0001. Significance tested by one-way ANOVA for B, C, and E left of divider. Significance tested by t-test for E right of divider.

Chapter 5. Conclusions and Future Directions

Congenital muscular dystrophies (CMDs) are severe disorders, having in common a pathogenic disruption of the cell membrane-extracellular matrix (ECM) interface.[65, 69] While the CMDs are candidates for recombinant adeno-associated virus (rAAV) gene therapy, the direct gene replacement strategy may not work due to the large size of the mutated genes in CMD, or due to the multiple loci at which a mutation may occur.[59, 60] As indicated by CMD pathology, the ECM is a functionally important part of tissue, in particular for the muscular and nervous systems.[13, 14] In this book, I developed a strategy of leveraging the ECM to create novel gene therapies for the CMDs. In each chapter, I employed the heparin binding (HB) domain of heparin-binding epidermal growth factor-like growth factor (HB-EGF) to accomplish different goals. The HB domain has strong binding affinity toward heparan-sulfate proteoglycans, which are a major part of the skeletal muscle ECM. [13, 281-283] Importantly, this ECM-targeting strategy may be modified to target different tissues and treat diseases other than the CMDs.

In Chapter 2, I developed a micro-laminin gene therapy for type 1A CMD. In this disease, mutations in the *LAMA2* gene, and loss of the laminin-α2 protein it encodes, disrupt the α-dystroglycan-laminin-211 connection, thereby allowing contracting myofibers to pull away from the ECM, whereupon they undergo apoptosis.[65] The *LAMA2* gene exceeds the rAAV packaging limit, making direct gene replacement with rAAV infeasible. I created the micro-laminin genes, *LAMA2(G1-5)* and *HB-LAMA2(G1-5)*, using segments of HB-EGF and the membrane-binding laminin-α2 globular (G) domains. Micro-laminins were expected to restore the cell membrane-ECM connection in the dy^W mouse model of MDC1A and reduce dystrophic pathology. rAAV-mediated expression of *HB-LAMA2(G1-5)* improved body weight-normalized forelimb grip strength in dy^W mice. Within treated muscle, myofibers expressing *HB-LAMA2(G1-5)* or *LAMA2(G1-5)* were on average larger, and displayed lower markers of pathology, myofiber size variation and central nucleation, than did non-expressing myofibers. Other important measurements of disease in dy^W mice, however, remained unchanged by the treatments. As such, it was clear that, in their current form, the micro-laminins I designed are only a starting point for this therapy strategy. Due to the modular way in which micro-laminins were created - the HB and laminin-α2 G1-5 domains are distinct functional units - they may be modified to improve their therapeutic efficacy. For instance, the HB domain may be swapped for another ECM-binding domain, such as a collagen-binding domain, if it proves more effective.[278, 279] Furthermore, the laminin-α2 G1-5 domain may be swapped for laminin-α2 G4-5 to specifically target dystroglycan.[43, 301, 307, 308] I created an *HB-LAMA2(G4-5)* gene using a natural cleavage site between G domains 3 and 4,[189, 308, 309] however this gene was not tested, herein. By using the *HB-LAMA2(G4-5)* gene, smaller than the *HB-LAMA2(G1-5)* gene by 1.328 kilobases, the ECM-binding domain could be further modified with, for example, a domain enabling micro-laminin polymerization. While it is currently unclear what domain could be used to enable micro-laminin polymerization, it is clear that polymerization is an important function of laminin-211 in addition to its membrane-ECM connecting function.[66-68]

In Chapter 3, I developed ECM-binding forms of two muscle trophic factors, insulin-like growth factor-1 (IGF-1) and follistatin (FST). Growth factor protein therapy is complicated by the need for repeated doses to maintain an effect, whereas growth factor gene therapy is vulnerable to adverse effects in off-target tissue. Using the HB domain, I created *HB-IGF1* and *HB-FST* genes, encoding growth factor proteins intended to have greater accumulation and retention in skeletal muscle than their native forms following rAAV-mediated expression. Indeed, HB-IGF1 and HB-FST proteins had ECM localization when expressed in skeletal muscle, different from their non-HB forms. The HB-proteins also appeared to accumulate to a greater degree in skeletal muscle than their non-HB forms. Trophic activity in skeletal muscle was either retained, as with HB-IGF1, or lost, as with HB-FST. While *HB-IGF1* expression resulted in increased muscle weight in dy^W mice following intramuscular (IM) injection, muscle weights were not improved following intravenous (IV) injection. Neither *IGF1*, *HB-IGF1*, nor multifunctional *HB-IGF1-LAMA2(G1-5)* expression improved total muscle pathology following IV injection in dy^W mice. An important question that remains unanswered in this book is whether HB-IGF1 and HB-FST proteins have less presence in the serum than IGF-1 and FST proteins following rAAV-mediated expression. Lower presence in the serum may indicate less risk of adverse effects in off-target tissues following gene therapy. As with *HB-LAMA2(G1-5)*, the HB-trophic factor genes may be modified either with different ECM-binding domains or with different trophic domains, in this case to target different diseases or tissues. Previous protein therapy studies, for instance, have fused the ECM-binding domain of placental growth factor-2 to various growth factors to target wound and bone defect repair.[279]

In Chapter 4, I developed a gene therapy for the α-dystroglycanopathies, a subgroup of the CMDs. In α-dystroglycanopathy, mutations in any one of 24 genes responsible for the complete glycosylation of α-dystroglycan disrupt the α-dystroglycan-laminin connection, resulting in a wide range of defects in the muscular and nervous systems.[83, 287, 289, 310] In order to cover all possible genetic causes of α-dystroglycanopathy, 24 unique gene replacement therapies would have to be designed - a long and expensive endeavor. I sought to create a universal gene therapy for the α-dystroglycanopathies by targeting the α-dystroglycan-ECM interface directly, rather than targeting a specific gene. I created the *HB-DAG1α* gene, encoding the HB domain fused to the C-terminus of α-dystroglycan. The HB-DAG1α protein was expected to bind both the ECM, and the transmembrane β-dystroglycan. I tested *HB-DAG1α* gene therapy in the Large[vls] mouse model of *LARGE*-related α-dystroglycanopathy. I found that while Large[vls] mice generally had a mild phenotype, they displayed marked central nucleation of skeletal myofibers. Furthermore, while *HB-DAG1α* expression did not affect other measurements of disease, it did reduce myofiber central nucleation. Recently studies have found that β-dystroglycan is translocated to the nucleus in certain cells, where it helps regulate nuclear functions and nuclear morphology.[37-39] That Large[vls] mice have such a high rate of central nucleation suggests that either the Large glycosyltransferase or, more likely, β-dystroglycan, has a role in myofiber central nucleation. If it is β-dystroglycan,

then this role is modified by connection to the ECM via α-dystroglycan. Furthermore,
that *HB-DAG1α* expression reduces central nucleation suggests that HB-DAG1α protein
is, at least in part, functioning like α-dystroglycan. As with micro-laminins, *HB-DAG1α*,
in its current form, represents only a starting point for this therapeutic strategy. It may be
modified with alternative ECM-binding domains. Additionally, the current α-
dystroglycan C-terminal domain also contains the mucin domain in addition to the region
that binds β-dystroglycan, and the mucin domain may not be required for this strategy. If
the mucin domain is removed, the shortened gene may be placed in a self-complementary
rAAV vector for more efficient expression in future studies.[311, 312]

Throughout this book, neonatal IV rAAV studies were complicated by low
transduction rates in skeletal muscle. This may be due to injection technique, loss of
rAAV genome during the early stage of disease, or an effect of adding a small volume of
Evans blue dye to the rAAV solution to visualize injection. Since CMDs are early onset
diseases, and this is reflected in the mouse models, the kinetics of rAAV-mediated
expression may have affected the transduction rate and treatment efficacy. There is a
kinetic lag of about one month between transduction and maximal gene expression.[191]
Furthermore, as seen in section 2.3.5, the expression of large proteins may take additional
time. During this time before potentially therapeutic protein expression, especially in dy^W
mice, dystrophy and loss of myofibers may include loss of vector genomes.